GCSE
Success

REVISION GUIDE

Physics

Carol Tear

Contents

Revised

The Electricity Supply

Energy Resources

People in modern societies use a lot of **energy resources**. Energy resources we use directly are called **primary energy resources**. Examples include diesel, petrol and solar energy. **Electricity** is a **secondary energy resource**. It has to be made from a primary resource.

We use a lot of electricity because:
• It's easy to transmit long distances.

• It can be used to do lots of different jobs.
• There is no pollution at the point where it is used.

A **sustainable energy** supply is one that meets our needs without leaving problems for future generations, for example using up all the resources, air pollution or causing climate change.

Efficiency

The efficiency of an energy transfer is the fraction, or percentage, of the energy input that is transferred to useful energy output.

Whenever energy is transferred from a primary resource to electricity some of it is wasted as heat in the process. No power station is **100 per cent efficient**.

In a **coal-fired** power station, for example, for every 100 J of energy stored in the coal that is burned, only 40 J is transferred to the electricity. Energy is a conserved quantity, which means that the total amount of energy remains the same. The 60 J of energy that is not transferred to the electricity is wasted and ends up as heat in the surrounding environment. This will cause a slight temperature increase in the surroundings.

$$\text{Efficiency} = \frac{\text{useful energy output}}{\text{total energy input}}$$

Or as a percentage:

$$\text{Efficiency} = \frac{\text{useful energy output}}{\text{total energy input}} \times 100\%$$

Energy transfers taking place in a process can be shown on a **Sankey energy flow diagram**.

The width of the arrows is proportional to the amount of energy represented by the arrow.

Energy transformation in a coal-fired power station

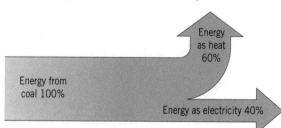

Energy as heat 60%

Energy from coal 100%

Energy as electricity 40%

A coal-fired power station

💡 Boost Your Memory

Draw two Sankey diagrams in your notes, one for a process that has more wasted energy than useful energy and one with more useful energy than wasted energy. Label the input energy, which is 100 per cent, and the useful energy and wasted energy. When you look at them, they will remind you of all the information summarised by a Sankey diagram.

Energy

Improving Efficiency

The European Union (EU) has banned the sale of 100 W filament lamps because about 90 per cent of the input energy is wasted as heat. Their efficiency is only 10 per cent because only 10 per cent of the output is useful light. Replacement lamps like compact fluorescent lamps (CFLs) are more efficient.

For example, a CFL that uses 18 J of energy, each second emits 10 J of light. Its efficiency is:

$$= \frac{10}{18} \times 100 = 56\%$$

A fluorescent lightbulb

Solar Energy

There are two ways of using solar energy:
- **Passive solar heating** uses radiation from the Sun to heat water.
- **Solar cells**, or **photocells**, transfer energy from sunlight into electricity.

Build Your Understanding

Solar cells contain crystals of silicon. Light gives the silicon atoms energy and knocks loose electrons from the atoms. The electrons flow as an electric current. The current is increased by increasing:
- The surface area that the light falls on.
- The light intensity.

❓ Test Yourself

1. Electricity is a secondary energy resource. Explain what this means.

2. A power station burns natural gas. Every second 755 MJ of electricity is generated from 1300 MJ of energy from the gas.

   a) Draw a Sankey diagram for the process.

   b) Calculate the efficiency of the power station.

3. An electric light uses 60 J of electrical energy every second and produces 8 J of light energy each second. Calculate its efficiency.

⭐ Stretch Yourself

1. A power station burns coal and converts 3500 kJ of chemical energy to electrical energy every second. 2010 kJ of energy is wasted as heat in the power station and 300 kJ is wasted in the overhead cables every second.

   a) Draw a Sankey diagram.

   b) Calculate the efficiency of the power station in delivering useful electricity to your home.

2. Explain why it is more efficient to use a primary energy resource to heat your home than to use an electric heater.

Generating Electricity

Power Stations

Turning a **generator** produces electricity. To turn the generators we connect them to **turbines.** We use different energy resources to turn the turbines. Wind and water flow can turn turbines directly. Steam is often used, produced by heating water. The heating is done by burning fuels, or using other heat sources. The diagram shows the parts of a **coal-fired** power station. In a modern gas-fired power station the hot exhaust gases from the burners are used to turn the turbines and then to heat water to steam which turns the turbines.

℗ Boost Your Memory

Learn to label the different parts of a power station and practice by drawing a flow diagram:
- furnace (or nuclear reactor)
- boiler
- turbine
- generator
- transformer

A coal-fired power station

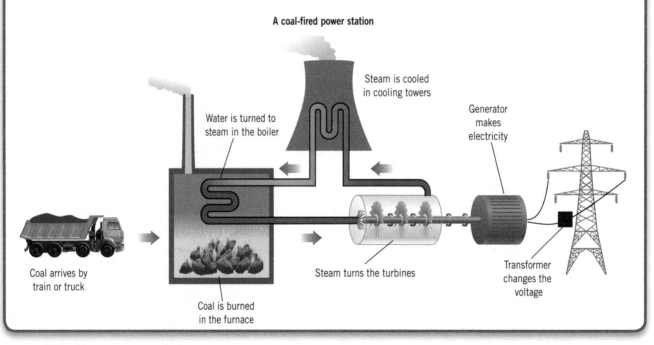

Steam is cooled in cooling towers

Generator makes electricity

Water is turned to steam in the boiler

Coal arrives by train or truck

Steam turns the turbines

Coal is burned in the furnace

Transformer changes the voltage

The Main Fuels

Electricity can be generated in large power stations from:
- **Fossil fuels** such as **coal**, **natural gas** and **oil** were formed over 300 million years ago. They will eventually run out. All the fossil fuels produce carbon dioxide when burned. Carbon dioxide absorbs infrared radiation and warms the atmosphere. The extra carbon dioxide from burning fossil fuels may be the cause of global warming and could cause climate change.

- In a **nuclear** reaction a large amount of energy is released from a small amount of **plutonium** or **uranium** (see page 13). One advantage is that no carbon dioxide is formed.
- **Biofuels** such as **wood, sugar, straw** and **manure** are materials from recently living plants and animals. Carbon dioxide is removed from the atmosphere when plants grow and then released when fuels are burned or fermented. As long as more plants are grown there is no net change. So biofuels are carbon neutral.

Energy

Lifecycle Assessments

When we decide which fuels to use in power stations we have to take lots of factors into account:

- Transport costs and the cost of the fuel.
- The availability of the fuel – will it run out or will it be difficult to get?
- The start-up time – how long it will take to commission (plan and build)?
- The cost of building the power station.
- How much it will cost to decommission (take apart at the end of its life) and to dispose of waste materials.
- Maintenance costs and availability of cooling water.
- Pollution and waste, for example whether carbon dioxide is emitted.
- The risk of accidents.

Generators

A voltage is **induced** across a coil of wire by moving a magnet into or out of a coil. Moving the coil instead of the magnet would have the same effect.

This effect is used in **dynamos** and **generators**. There are three ways to increase the voltage:
- Use stronger magnets.
- Use more turns of wire in the coil.
- Move the magnet (or the coil) faster.

This process is called **electromagnetic induction**.

A generator in a power station uses an electromagnet to produce a magnetic field. The electromagnet rotates inside coils of wire so that the coils are in a changing magnetic field and a voltage is induced.

A voltage is induced when there is movement

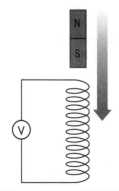

Build Your Understanding

A voltage is only induced when there is movement of the magnet or coil. The direction of the voltage is reversed when the movement is reversed, or when the poles of the magnet are changed.

Another way to increase the voltage is to place an iron core inside the electromagnet coil.

A generator

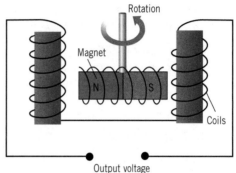

Output voltage

✓ Maximise Your Marks

Make sure you can compare advantages and disadvantages of the different fuels on pages 6–7 and the renewables in the next section.

❓ Test Yourself

1. What is the difference between the energy source in a coal-fired power station and a nuclear power station?

2. What is the energy source in a nuclear power station?

3. Which of these are biofuels: chicken manure, coal, natural gas, uranium, willow?

⭐ Stretch Yourself

1. Should we build more nuclear or fossil fuel power stations? Give two reasons for and two reasons against each type.

2. Give two advantages and two disadvantages of a biofuel power station.

Renewable Sources of Energy

Renewable Resources

Renewable sources of energy are those that are being made today and so will not be used up. Most renewable resources make use of the Sun's energy.

The Sun evaporates water and causes the rain that fills the rivers. The Sun also causes convection currents that produce winds, which produce waves. The exceptions are **geothermal** heat which originates from inside the Earth and the tides which are caused by movement of the Moon.

These are the renewable resources:

- **Hydroelectric Power (HEP)** is electricity generated using fast flowing water to turn the turbines. Dams are built to form reservoirs of water for the power stations.
- **Wind turbines** use the energy of the wind to turn the turbines. Wind farms are collections of wind turbines.
- **Solar cells** produce electricity by the action of sunlight on the material inside the cell. Solar panels heat water for heating and washing.
- **Wave generators** use the movement of waves to generate electricity.
- **Biofuels**, or **biomass** fuels (see page 6), can be burned, or they can be used to produce methane gas or alcohol.
- **Geothermal energy** is used in power stations in places where the Earth's crust is thin and the heat is close to the surface. Geothermal power stations can use this heat.
- **Tides** change the height of the sea in some areas so much that it is worth using it to generate electricity.

💡 Boost Your Memory

To remember all the renewables make up a mnemonic that contains the first letters of each. For example: **H**igh **W**inds **S**ometimes **W**ave **B**ig **G**reen **T**rees.

Advantages and Disadvantages

Source	Advantages	Disadvantages
HEP	No waste or air pollution. No fuel cost. Can generate a lot of electricity in mountainous areas.	Rainfall or snow is not constant. Building dams and flooding valleys changes the environment. Homes, farmland and natural habitats are lost forever.
Wind	No polluting waste gases. The wind is free, so the cost of electricity is low.	The wind does not always blow. Some people consider wind turbines noisy and an eyesore, and they take up a lot of space for the amount of electricity generated.
Solar cells	No air pollution. No fuel cost. No moving parts so they do not need much maintenance. They have a long life. They can be used in remote locations.	Cannot produce power at night or in bad weather.
Wave generators	No air pollution. No fuel cost. Lots of available energy.	Waves are destructive, so it has been difficult to develop the technology.

Advantages and Disadvantages (cont.)

Source	Advantages	Disadvantages
Biofuels	A way of disposing of waste. In the case of waste and manure, the pollution would be produced anyway. Carbon neutral.	Power stations need a steady supply. They produce pollution – carbon dioxide, other gases and ashes.
Geothermal	No air pollution. No fuel cost.	Earthquakes and volcanic action may damage the power station.
Tides	The tide does not depend on the weather.	The habitats of many birds and other animals may be destroyed.
Solar panels	No air pollution or fuel costs.	They do not work so well in poor weather.

✓ Maximise Your Marks

Make sure you can explain why it is important to have a reliable supply of electricity and why this means it is best to have a mixture of ways of generating electricity and not have to rely on one method.

❓ Test Yourself

1. Explain how a hydroelectric power station works.

2. Choose two renewable energy sources that are the most suitable for use in the UK. Give your reasons for choosing them.

3. Choose two renewable energy sources that are not very suitable for use in the UK. Give your reasons for deciding they are not suitable.

★ Stretch Yourself

1. Jenny says that it would be best to generate all our electricity from wind farms and wave generators.

   a) List some advantages of wind farms and wave generators.

   b) Explain some reasons why Jenny's idea is not a good one.

2. Explain which energy source is most suitable for:

   a) Providing an electric light for a rural bus stop.

   b) Providing energy to mine and process aluminium ore in a mountainous region.

Electrical Energy and Power

Energy and Power

We use electrical appliances at home to transfer energy from the mains supply to:
- heating
- light
- movement and sound.

In two hours an electric lamp uses twice as much energy as it uses in one hour. The **power** of an electrical appliance tells us how much electrical energy it transfers in a second.

Power, P is measured in **watts** (W) where:

1 W = 1 J/s (joule/second).

Appliances used for heating have a much higher rating than those used to produce light or sound.

The amount of energy transferred from the mains appliance depends on the power rating of the appliance and the time for which it is switched on. **Energy** transferred from electricity is worked out by:

Energy = power × time

$$E = P \times t$$

Energy, E is measured in:
- **joules** (J) when the power is in watts and the time, t, is in seconds.
- **kilowatt hours** (kWh) when the power is in kilowatts and the time, t, is in hours.

Example: A 300 W electric pump is switched on for one minute. The energy used is:

$$E = 300 \text{ W} \times 60 \text{ s}$$

$$E = 18000 \text{ J}$$

Example: A 2.2 kW kettle used for 10 minutes:

$$E = 2.2 \times \frac{10}{60} = 0.37 \text{ kWh}$$

Power ratings of some electrical appliances

| 2 kW | 1 kW | 800 W |

Build Your Understanding

The power used by an electrical appliance is the rate at which it transfers electrical energy. A 100 W light bulb uses more electrical energy than a 60 W light bulb every second. It transfers more energy per second.

Example: A 100 W lightbulb uses 6 kJ of energy

in $t = \dfrac{6000 \text{ J}}{100 \text{ W}} = 60 \text{ s}$

A 60 W lightbulb uses 6 kJ in $t = \dfrac{6000 \text{ J}}{60 \text{ W}} = 100 \text{ s}$

A standard lightbulb

Energy

Paying for Electricity

The units on an electricity bill, and measured by an electricity meter, are kilowatt hours. The cost of each unit of electricity varies. The electricity bill is calculated by working out the number of units used and multiplying by the cost of a unit.

Cost of electrical energy used =

power in kW × time in hours × cost of one unit
Or

Cost = number of kW h used × cost of one unit

Example: The 800 W toaster in the diagram is used for half an hour and the cost of a unit is 12p:

Cost = 0.8 kW × 0.5 hours × 12 p/kW h

Cost = 4.8p = 5p to nearest 1p

✓ Maximise Your Marks

Remember that the kilowatt hour is a unit of energy – not power. Power is measured in watts or kilowatts.

Power tells you how quickly energy is being used. A 3 kW fire uses the same energy in one hour as a 1 kW fire does in three hours. The energy is power × time.

3 kW fire: 3 kW × 1 h = 3 kWh

1 kW fire: 1 kW × 3 h = 3 kWh

The unit is kilowatts multiplied by hours = kWh.

A common mistake is to say kilowatts per hour kW/h.

Power, Current and Voltage

The mains voltage in the UK is 230 V. Electrical power depends on the current and the voltage:

Power = current × voltage

$P = I \times V$

Power is measured in **watts** (W), current, I, in **amps** (A) and voltage, V, in **volts** (V).

A torch with a 3.0 V battery has a current of 0.3 A. Its power is:

$P = 3.0 \times 0.3 = 0.9$ W

Build Your Understanding

To calculate the current in this 1.3 kW hairdryer rearrange the equation to give:

$$I = \frac{P}{V}$$

$$\frac{1300 \text{ W}}{230 \text{ V}} = 5.65 \text{ A}$$

A hairdryer

✓ Maximise Your Marks

Be careful with time when doing calculations. Change minutes to seconds if you are using energy in joules and to hours if the energy is in kilowatt hours.

❓ Test Yourself

1. What unit is power measured in?

2. What is the power of a 230 V lamp with a current of 0.05 A?

3. Referring to the figure on page 10 how much energy in kWh does the kettle in the diagram use in six minutes?

4. Referring to the figure on page 10 how much energy in joules does the toaster in the diagram use in one minute?

⭐ Stretch Yourself

1. What is the current in the microwave oven in the diagram?

2. A unit costs 12p. How much does it cost to use the microwave oven for 15 minutes?

Electricity Matters

The National Grid

In the UK the **National Grid** is a network that connects all the generators of electricity, like power stations, to all the users, for example homes and workplaces.

Mains electricity is an **alternating current (a.c.)** which means that the **current** keeps changing direction. Current from batteries is **direct current (d.c.)** meaning it always flows in the same direction.

Advantages of having a National Grid are:
• Power stations can be built where the fuel reserves are, or near the sea or rivers for cooling.
• Pollution can be kept away from cities.
• Power can be diverted to where it is needed, if there is high demand or a breakdown.

A disadvantage of the National Grid is that power is wasted heating the power lines.

The UK National Grid

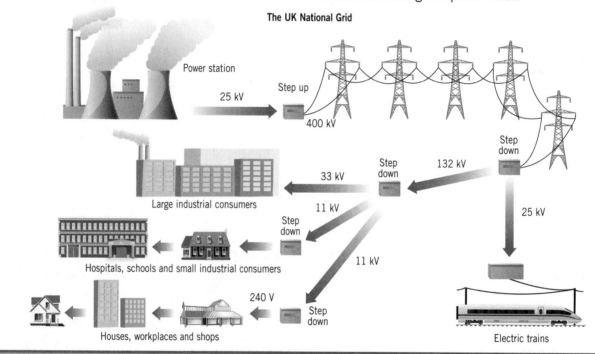

Power station — 25 kV — Step up — 400 kV — Step down — 132 kV — Step down — 33 kV — Large industrial consumers — 11 kV — Step down — Hospitals, schools and small industrial consumers — 11 kV — 240 V — Step down — Houses, workplaces and shops — Step down — 25 kV — Electric trains

Transformers

A **transformer** changes the size of an **alternating voltage.** Transformers will not work with a constant voltage. One of the reasons we have an a.c. mains supply is that the voltage is alternating and can be changed using transformers. This means it can be distributed more efficiently by using high voltages, for example 400 kV (1 kV = 1000 V).

Step-up transformers increase voltage and **step-down transformers** decrease voltage.

The voltage is stepped-up at the power station, transmitted at high voltage, to reduce power losses and stepped down at the local sub-station.

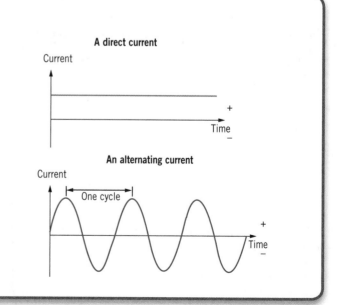

A direct current

Current

Time

An alternating current

Current

One cycle

Time

Build Your Understanding

Example: To supply 100 kW of power we can use:

High voltage and small current:
$P = 100 \text{ kW} = 1 \text{ A} \times 100 \text{ kV}$

Low voltage and large current:
$P = 100 \text{ kW} = 100 \text{ A} \times 1 \text{ kV}$

The heating effect in the cables depends on the current. By making the current as small as possible the energy wasted as heat in the cables is reduced. The current can be small if the voltage is high. There are many hundreds of miles of overhead power cables so this saves a lot of energy.

Nuclear Power Stations

Nuclear power stations are like fossil fuel power stations, but instead of burning fuel they use a nuclear reaction, **nuclear fission**, to transfer energy as heat. Nuclear fuels are **uranium** and **plutonium**, which are radioactive. Uranium is mined. Plutonium is formed in nuclear reactors. A disadvantage is that **radioactive waste** is produced that remains dangerous to living things for millions of years.

Radioactive materials emit **ionising radiation**. This is dangerous to living things (see page 36).

The Nuclear Option

Advantages of nuclear power stations include:
- No carbon dioxide is formed in the nuclear reaction.
- A small amount of fuel releases a large amount of energy.
- The fuel is not expensive, so the running costs of nuclear power stations are not high.

Disadvantages include:
- There is the risk of an accidental emission of radioactive material while the power station is operating.

- Accidental emissions could happen due to human error, for example at Chernobyl in Ukraine, or to natural disasters, like earthquakes, for example at Fukushima in Japan.
- Both the power station and the fuel are targets for terrorists.
- The risks to living things from radioactive materials mean that there are high maintenance costs and high decommissioning costs. Radioactive waste must be stored safely for thousands of years.

✓ Maximise Your Marks

Know how to make a case both for and against building nuclear power stations.

💡 Boost Your Memory

List and learn the similarities and differences between fossil fuel (page 6) and nuclear power stations.

❓ Test Yourself

1. What is the difference between a.c. and d.c. electricity?
2. What is the National Grid?
3. Give two similarities and two differences between a nuclear power station and a coal-fired power station.
4. Why is radioactive waste dangerous?

⭐ Stretch Yourself

1. What device is used inside a mobile phone charger to convert the 230 V supply to 12 V?
2. Why does the National Grid transmit electricity at 400 kV?

Energy

Particles and Heat Transfer

Energy

Specific Heat Capacity

When the temperature of an object increases it has gained energy. The amount of energy depends on:
- The temperature change, θ
- The mass of the object, m
- The specific heat capacity, c

The **specific heat capacity** is different for different materials. It is the energy needed to increase the temperature of 1 kg of the material by 1°C and is measured in J/kg °C.

Energy = mass × specific heat capacity × temperature change

$$E = m \times c \times \theta$$

Build Your Understanding

This is how to measure the specific heat capacity of a metal block:
- Measure the temperature and the mass of the block, m.
- Use an electric heater to raise the temperature of a metal block. Energy supplied, **E = power × time**.
- Measure the temperature of the block at the end of the heating time and calculate the increase in temperature θ.
- Calculate the specific heat capacity of the metal,

$$c = \frac{E}{m \times \theta}$$

Heat Transfer

Heat can be transferred by **conduction**, **convection** and **radiation**. It is only transferred from hotter things to cooler things.

In a hot solid, particles vibrate more. They collide with the particles next to them and set them vibrating. The kinetic energy is transferred from particle to particle. Metals are the best conductors. Solids are better than liquids. Gases are very poor **conductors**. They are **insulators**.

In a hot fluid (gas or liquid) the particles have more kinetic energy so they move more. They spread out and the fluid becomes less dense. The hot fluid rises above the denser cold fluid forming a **convection current**.

All objects emit and absorb **infrared radiation**. The higher the temperature the more they emit.

When objects absorb this energy their temperature increases.

Radiation will travel through a vacuum – it does not need a medium (material) to pass through:
- Dark and matt surfaces are good absorbers and emitters of infrared radiation.
- Light and shiny surfaces are poor absorbers and emitters of infrared radiation.
- Light and shiny surfaces are good reflectors of infrared radiation.

Convection currents

Conduction in a solid – energy is transferred from molecule to molecule

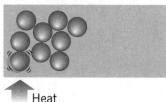

Heat

✓ Maximise Your Marks

Students sometimes lose marks by saying that 'heat rises'. This is not true. It is hot gas or liquid that rises. Sometimes people say that cold is coming into a house. It's better to say warm air is getting out. The cold air only moves to replace the warm air.

Insulation

When we insulate our homes we reduce the heat lost, we use less fuel and it costs less.

Still air is a good insulator, so materials with air trapped in them are often used:
- In cavity walls the air gap between the walls stops conduction.
- In cavity wall insulation the cavity is filled with foam or mineral wool.
- Loft insulation using layers of fibreglass or mineral wool
- Reflective foil on walls reflects infrared radiation.
- Draught-proofing stops hot air leaving and cold air entering the house.

All these improvements cost money to buy and install, but they save money on fuel costs. You can work out the **payback time** which is the time it takes before the money spent on improvements is balanced by the fuel savings, and you begin to save money:

$$\text{Payback time (in years)} = \frac{\text{cost of insulation}}{\text{cost of fuel saved each year}}$$

If the price of the fuel increases, the payback time will be less.

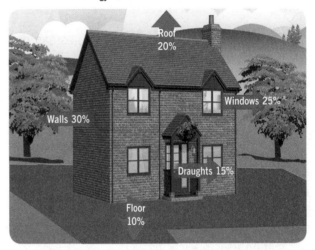

Energy flow from an uninsulated house

Roof 20%
Windows 25%
Walls 30%
Draughts 15%
Floor 10%

Build Your Understanding

Metals are made of a lattice of positive ions and 'free' electrons that can move through the lattice.

This makes them good conductors of heat because they have 'free' electrons to carry the energy. Because electrons are negatively charged this also makes metals good conductors of electricity.

Heat is transferred to, or from, an object at a rate that depends on:
- Its surface area and volume.
- The material it's made from.
- What the surface is like.
- The temperature difference between the object and its surroundings.

The bigger the temperature difference, the faster the object will heat up or cool down.

? Test Yourself

1. How much energy is needed to raise the temperature of 1 kg of water by 2°C? (water, c = 4200 J/kg°C)

2. Draw and label convection currents in a beaker of water heated by a Bunsen burner.

3. A new boiler costs £2000. It saves £100 on fuel costs each year. What is the payback time?

★ Stretch Yourself

1. 4400 J of energy raises the temperature of 0.5 kg of aluminium from 15°C to 25°C. Calculate the specific heat capacity of aluminium.

2. How does cavity wall insulation work?

Practice Questions

 Complete these exam-style questions to test your understanding. Check your answers on pages 120–21. You may wish to answer these questions on a separate piece of paper.

1 A power station burns coal and converts 3.5 MJ of chemical energy to 1.19 MJ of electrical energy every second. The efficiency of the power station is: (1)

A 2.31%

B 29%

C 34 %

D 47%

2 The energy wasted in electrical cables can be reduced by transmitting energy through cables with: (1)

A A high current.

B A high heat capacity.

C A high resistance.

D A high voltage.

3 In power stations turbines are used to turn generators to produce electricity. A nuclear power station uses nuclear fission.

a) What fuel is used in a nuclear power station? (1)

b) What is meant by nuclear fission? (1)

c) How does a nuclear power station use the heat produced by nuclear fission? (2)

d) Describe how a hydroelectric power station generates electricity differently from a nuclear power station. (1)

e) Explain one advantage and one disadvantage of a hydroelectric power station compared to a nuclear power station. (2)

4 A combined cycle gas turbine (CCGT) power station burns gas and uses the hot exhaust gases to turn the turbines. The hot exhaust gases are then used to heat water to steam to turn more turbines. Every second 755 MJ of electricity is generated from 1300 MJ of chemical energy in the gas.

a) Use the equation (electrical energy output/fuel energy input × 100%) to calculate the efficiency of the power station. (2)

b) Explain two ways a CCGT power station is different from a coal-fired power station. (2)

c) Explain one disadvantage of using gas to fuel a power station. (1)

5 Ian decides to try and cut down the amount of electricity he is using so he keeps some notes for a week:

Appliance	Power rating (kW)	Time used each day (hours)
Electric fire	3	2
Electric kettle	1.2	0.5
Electric light	0.1	5
Computer	0.3	4
Television	0.5	3

a) How many units does the electric fire use in a day? (1)

b) Which appliance is used for the most time? (1)

c) How much does it cost to use the fire for a day? Electricity costs 10p per unit. (1)

6 Explain the advantages and disadvantages of a coal-fired powered station compared to a wind farm. In your answer, aim to write between 8 and 12 lines. (6)

How well did you do?

| 0–6 | Try again | 7–13 | Getting there | 14–18 | Good work | 19–23 | Excellent! |

Describing Waves

Transverse and Longitudinal Waves

A **wave** is a **vibration** or disturbance transmitted through a material (a medium) or through space. Waves transfer energy and information from one place to another, but they do not transfer material.

A **transverse wave** has vibrations at right angles (perpendicular) to the direction of travel. The wave has **crests** and **troughs**.

Examples include water waves, waves on strings or rope, light and other electromagnetic waves.

A **longitudinal wave** has vibrations parallel to the direction of wave travel. It has **compressions** and

between these are stretched parts called **rarefactions**.

Examples include sound waves, ultrasound waves, which are sound waves with frequency greater than 20 kHz, and waves along a spring.

A longitudinal wave

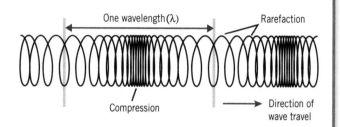

A transverse wave

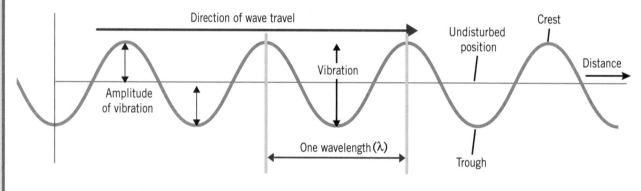

Build Your Understanding

Longitudinal waves travel through solids, liquids and gases, but they cannot travel through a vacuum.

✓ Maximise Your Marks

Use words like the ones below to explain what you mean, so the examiner can award you the marks:
- Oscillation for side to side, up and down, or back and forth movements.
- Perpendicular for at right angles.
- Parallel for in the same direction.

Transverse electromagnetic waves can travel through a vacuum. They are oscillations of a magnetic and electric field.

There are two types of seismic waves (see page 22) called P-waves and S-waves:
- P-waves are longitudinal waves that travel through solid or liquid rock. They travel faster than S-Waves.
- S-waves are transverse waves that can only travel through solid materials in the Earth.

Wave Properties

Amplitude is the maximum displacement (change in position) from the undisturbed position.

Wavelength (λ) is the distance in metres from any point on the wave to an exactly similar point.

Frequency is the number of waves that pass a point in one second. This depends on how fast the source of the waves is vibrating. The frequency is usually expressed in hertz (Hz) where one Hz is one cycle (wave) per second.

Example: Four waves pass a point in one second. The frequency = 4 Hz.

Example: A wave takes two seconds to pass a point. The frequency = 0.5 Hz

Wave speed depends on the medium that the wave is travelling through:

distance wave travels = wave speed × time ($d = v \times t$)

Example: A water wave travels at 5 cm/s. In 4 s it travels d = 5 cm/s × 4 s = 20 cm.

The **wave equation** relates the wavelength and frequency to the wave speed. For all waves:

wave speed = frequency × wavelength ($v = f\lambda$ where f is in Hz, λ is in m, v is in m/s).

Example: Water waves with wavelength λ = 10 cm and frequency f = 4 Hz have a speed v = 10 cm × 4 Hz= 40 cm/s.

Waves with different frequency and wavelength

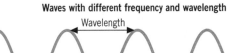
Wavelength

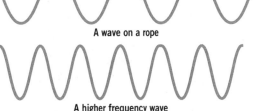

A wave on a rope

A higher frequency wave

Build Your Understanding

The wavelength of light is very small and the wave speed and frequency are very high.

Example: Light travels at 300 000 000 m/s = 3×10^8 m/s.

Wave speed $v = 3 \times 10^8$ m/s.

Green light has wavelength, λ = 520 nm = 5.2 $\times 10^{-7}$ m (1 nm is 1 nanometre = 1×10^{-9}m).

To calculate the frequency, f, of the light waves, rearrange the wave equation: $f = \dfrac{v}{\lambda}$

$$f = \frac{3 \times 10^8 \text{ m/s}}{5.2 \times 10^{-7}\text{m}} = 5.8 \times 10^{14} \text{ Hz}$$

Sound Waves

For sound and ultrasound waves:
- The **pitch** of the sound is the frequency of the vibrations.
- The **loudness** of the sound depends on the amplitude of the vibrations

Sound travels much faster in solids than in liquids and faster in liquids than in gases.

✓ Maximise Your Marks

A common mistake is to mark the amplitude from the top of a peak to the bottom of a trough – this is twice the amplitude.

? Test Yourself

1. Draw a transverse wave and mark the amplitude and the wavelength.

2. How is a longitudinal wave different from a transverse wave?

3. A sound wave has frequency 256 Hz and wavelength 1.30 m. Calculate the speed of the sound.

★ Stretch Yourself

1. Water waves pass a marker at a rate of one every four seconds. What is their frequency?

2. What is the frequency of radio waves with a wavelength of 3 m?

Wave Behaviour

Reflection

All waves can be reflected, but this does not prove they are waves because particles also show these effects. According to the **law of reflection**, the angle of incidence equals the angle of reflection. This can be shown on a wave diagram or a ray diagram. **Echoes** are reflections of sound waves from hard surfaces.

Reflection

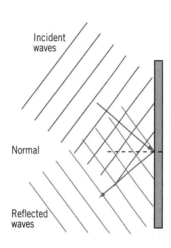

Incident waves

Normal

Reflected waves

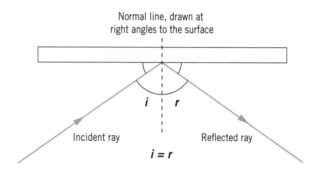

Normal line, drawn at right angles to the surface

i r

Incident ray

Reflected ray

$i = r$

Refraction

When waves enter a different **medium** they change **speed**. The wavelength changes, but the frequency stays the same. They change direction unless they are travelling along the normal to the boundary. This is called **refraction**.

As light waves enter a **denser medium** they slow down and bend towards the normal. As light waves enter a **less dense medium** they speed up and bend away from the normal. Water waves slow down as they go from deep water to shallow water.

Refraction of light in a glass block

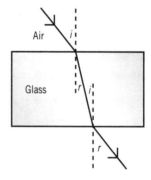

Air

i

Glass

r i

r

Refraction causes waves to change direction

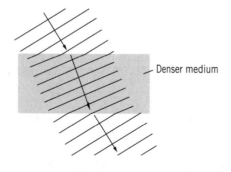

Denser medium

Waves

Build Your Understanding

The image in a plane (flat) mirror is the same distance behind the mirror as the object is in front. It is a virtual image. The image is also upright and laterally inverted.

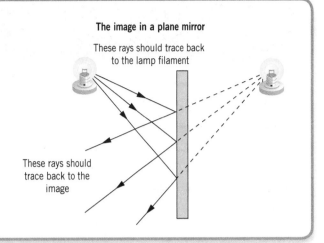

The image in a plane mirror

These rays should trace back to the lamp filament

These rays should trace back to the image

Diffraction

Diffraction is the spreading out of a wave when it passes through a gap. The effect is most noticeable when the gap is the same size as the wavelength. Particles cannot be diffracted. So diffraction is good evidence for a wave.

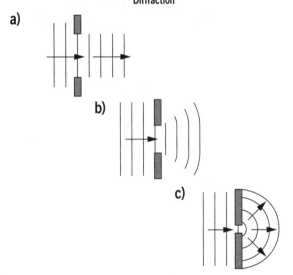

Diffraction

a)

b)

c)

Build Your Understanding

The wavelength of sound waves is about a metre, so when sound waves are diffracted through doorways people can hear round corners.

Water waves are sometimes diffracted by harbour entrances.

The wavelength of light is about half of a millionth of a metre – so small that diffraction effects are only noticeable when light waves pass through gaps of about a hair's width. Larger gaps have shadows with sharp edges. This is one reason why it took so long for scientists to realise that light can behave as a wave.

✓ Maximise Your Marks

It is important to compare the size of the gap with the wavelength when you are explaining whether diffraction will occur and when the effect will be biggest.

? Test Yourself

1. If the angle of incidence is 30°, what is the angle of reflection?

2. A ray goes from air to glass with an angle of incidence of 30°. Is the angle of refraction bigger, smaller or the same?

3. Draw a ray diagram to show light at an angle to the normal crossing a boundary from a) air to glass and b) from water to air.

4. Draw a ray diagram to show how two mirrors can be used to see round a corner.

★ Stretch Yourself

1. Sketch a diagram of:

   a) Water waves with wavelength 1 cm passing through a 1 cm gap.

   b) Light waves with wavelength 5×10^{-7} m passing through a 1 mm gap.

   c) The same light waves passing through a 1 micrometre gap $(1 \times 10^{-6}$ m).

Seismic Waves and the Earth

Seismic Waves

Seismic waves are caused by earthquakes or large explosions, such as quarrying operations or atomic bomb tests. P-waves and S-waves both travel *through* the Earth and are detected at monitoring stations all around the Earth by instruments called **seismometers.** They record the waves arriving on a **seismograph.**

P-waves travel faster through the Earth than other seismic waves, so they are the first to be detected after an earthquake. They are primary waves and are longitudinal.

S-waves are detected after primary waves.

The time delay between the arrival of the P-wave and S-waves is greater at stations further from the earthquake, so the distance to the earthquake centre can be calculated. Using information from several stations the position of the earthquake is worked out.

The structure of the Earth's crust is investigated by setting up monitoring equipment at different points and setting off a controlled explosion. The waves are then recorded arriving at the monitoring points. Analysing this data gives information about the rock structure.

A seismograph

P   S

| | | | | |
| 0 | 1 | 2 | 3 | 4 | 5 |
Earthquake   Time/minutes
occurs

A seismometer

Build Your Understanding

Data from all the monitoring stations after earthquakes, shows a shadow zone on the opposite side of the Earth to the earthquake where no S-waves are detected.

This can be explained by the Earth having a liquid core because S-waves don't travel through liquids.

The diagram below shows the paths of P-waves and S-waves through the Earth following an earthquake. The waves are reflected and refracted at the boundaries.

P- and S-waves travel through the Earth

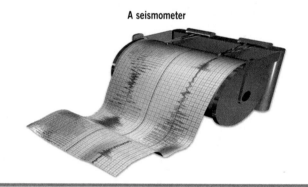

Epicentre

P- and S-waves   P- and S-waves

P-waves only

ⓘ Boost Your Memory

Think of P-waves as primary waves and S-waves as secondary waves. After an earthquake, P-waves arrive first and S-waves arrive second. P-waves have a faster speed.

Waves

The Structure of the Earth

The Earth is made up of a **core**, a **mantle** and a **crust**. The theory of continental drift was first suggested by Alfred Wegener in 1912, based on the way the continents appeared to fit together like a jigsaw, and matching fossils and rock layers on different continents. Due to lack of evidence, it was not until 1967 that the theory of plate tectonics was accepted.

The Earth's crust consists of a number of moving sections called **tectonic plates**. The mantle behaves like a very thick liquid. The plates move because of convection currents in the mantle.

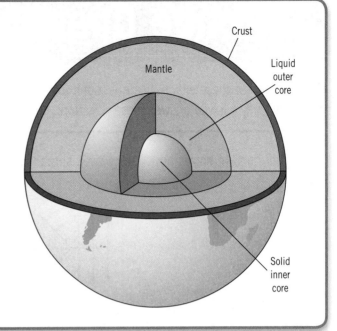

Crust

Mantle

Liquid outer core

Solid inner core

Build Your Understanding

The Earth must be older than the oldest rocks which are about four thousand million years old. Rock processes seen today can explain changes that happened in the past. Rocks are worn down by erosion. Sediments are deposited in layers by rivers and seas and form new rocks. All the continents would be worn flat if new mountains were not being formed.

✓ Maximise Your Marks

Be able to explain that Wegener's theory was not believed because:
- He was not a geologist.
- There was no explanation of how the continents move – no mechanism.
- There was no way of measuring any movement – no evidence.

The Rock Cycle

Magma flows out of the mid-ocean ridges forming new rock, so the sea floor spreads by a few centimetres a year and this causes the continents to move apart.

At boundaries, where two plates collide, rocks are pushed up forming new mountain ranges or when plates slide past each other this can sometimes cause earthquakes. At boundaries where magma comes to the surface there are volcanoes.

This is why earthquakes, mountains and volcanoes are generally found along the edges of the tectonic plates.

? Test Yourself

1. Which arrives first after an earthquake, P-waves or S-waves?

2. What is a seismometer?

3. What does the theory of plate tectonics suggest about the surface of the Earth?

4. How can the distance to an earthquake be worked out from seismometer readings?

★ Stretch Yourself

1. Are a) P-waves and b) S-waves transverse or longitudinal?

2. How do scientists know there is a liquid part of the Earth's core?

Practice Questions

 Complete these exam-style questions to test your understanding. Check your answers on page 121. You may wish to answer these questions on a separate piece of paper.

1 The diagram shows how water waves spread out after passing through a gap. (1)

This effect is called:

A Diffraction

B Dispersion

C Reflection

D Refraction

2 Which of these is **not** evidence for the theory that the continents of Europe and North America are drifting apart? (1)

A There are similar fossils in the two continents, but different modern animals.

B The shape of the continents suggest that they fit like pieces of a jigsaw.

C The rocks at either side of the Mid-Atlantic ridge are in strips that are magnetised in opposite directions.

D Mountains, like the Himalayas, are formed when a continental plate collides with another continental plate.

3 Use some of the words below to complete the sentences. (2)

disturbance   energy   longitudinal   matter   medium   transverse   vibrates

A wave is the movement of a _____ through a _____ . Waves transfer _____ but not _____ . A wave is caused by something that _____ .

4 Give two differences between earthquake waves called P-waves and S-waves. (2)

....

....

5 Some ocean waves have a wavelength of 140 m and one wave passes a rock every 20 seconds. The distance from the top of a crest to the bottom of a trough is 3 m. Tick (✓) **true** or **false** for each statement. (1)

	True	False
The frequency of the waves is 20 Hz.		
The wave speed is 7 m/s.		
The amplitude is 3 m.		

6 When water waves move from deep water to shallow water they are refracted. Explain what this means. (1)

....

....

7 Some ocean waves have a wavelength of 200 m and one wave passes a rock every 10 seconds.

a) What is the frequency of the waves? (1)

....

b) Write down the equation you can use to work out wave speed from wavelength and frequency. (1)

....

c) What is the speed of the waves? (1)

....

8 This diagram shows a seismograph from a place near to an earthquake.

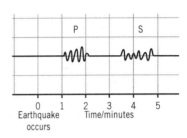

a) Which type of wave travels faster, P-waves or S-waves? Explain how you can tell. (2)

....

b) Explain how the seismograph from a place a few hundred kilometres further away from the earthquake would look different. (2)

....

....

c) Explain how the seismograph from a place on exactly the opposite side of the Earth would look different. (2)

....

....

9 Explain the similarities and differences between sound waves and light waves.
(In your answer, aim to write between 8 and 12 lines.) (6)

....

....

....

....

....

....

....

....

How well did you do?

| 0–6 | Try again | 7–12 | Getting there | 13–17 | Good work | 18–23 | Excellent! |

The Electromagnetic Spectrum

The Electromagnetic Spectrum

The spectrum of electromagnetic waves is continuous from the longest wavelengths (**radio waves**) through to the shortest wavelengths (**gamma rays**).

All electromagnetic waves are transverse waves that can travel through a vacuum. They all travel through empty space at a speed of 300 000 000 m/s.

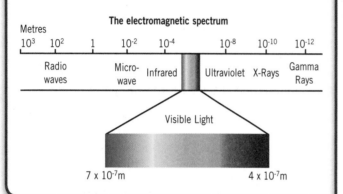

Radio waves have the longest wavelength, lowest frequency and lowest energy. Gamma rays have the shortest wavelength, highest frequency and highest energy. The wavelengths are related to the frequencies using the wave equation (see page 19).

⚡ Boost Your Memory

The spectrum goes from radio waves to gamma rays:
- Remember the colours of visible light using the mnemonic **R**ichard **O**f **Y**ork **G**ave **B**attle **I**n **V**ain (ROY G BIV).
- Next remember infrared is under red and ultraviolet is above violet in energy.
- Then remember waves (radio and micro) are below infrared and rays (X and gamma) are above ultraviolet.

Build Your Understanding

The wavelength of visible light is about half a thousandth of a millimetre which is so small that it is not obvious that it is a wave. Sometimes light behaves as a stream of particles.

There are not fixed boundaries between the different types of electromagnetic waves. The diagram above shows the typical values for each type.

Gamma rays are always shown as shorter wavelength than X-rays, but the ranges overlap. The real difference is that gamma rays come from radioactive materials and X-rays are produced in an X-ray tube.

✓ Maximise Your Marks

Remember these typical wavelengths:
- 1 km = radio
- 1 cm = microwave
- 0.1 mm = infrared
- 10^{-10} m = X-ray

Radiation

Radiation is **transmitted** from the source to the detector. Some materials **absorb** some types of electromagnetic radiation. The further the radiation travels through an absorbent material the lower its intensity will be when it reaches the end of its journey. More of the radiation energy is absorbed by the material and it heats up.

On some journeys the radiation is **reflected** at a boundary between two different materials.

When the energy of the radiation is absorbed by the detector it may:
- Have a heating effect.
- Produce small electric currents in aerials (if microwaves or radio waves).
- Make chemical reactions more likely, for example light causes photosynthesis in plants.
- Ionise atoms (if ultraviolet, X-ray or gamma).

Energy and Intensity

When electromagnetic radiation strikes a surface, the **intensity** of the radiation is the amount of energy arriving at a square metre of the surface each second.

Radiation is emitted from a source and travels towards a destination. On this journey the radiation spreads out, so the further away a detector is from the source the less energy is detected. The intensity of the radiation can be increased by moving closer to the source.

Intensity can also be increased by increasing the radiation from the source.

Radiation spreads out from the source

Spotlight

A1

Larger area
less intensity

A2

The identical spotlights on the same area double the intensity

 Spotlight   Spotlight

Double intensity

A

Build Your Understanding

To compare the intensity of radiation we measure the amount of energy falling on one square metre of the surface in each second.

The energy absorbed by the surface or detector can be calculated by:

Energy (J) = intensity (W/m^2) × time (s)

Electromagnetic radiation sometimes behaves as waves, and sometimes as packets of energy called **photons.** A gamma ray photon has the most energy. The energy of the photons increases with the frequency of the radiation.

Blue light has more energy than red light

Spotlight   Spotlight

A   A
Higher energy   Lower energy

? Test Yourself

1 State two properties that all electromagnetic waves have in common.

2 State one difference between infrared and ultraviolet rays.

3 Electromagnetic radiation has a wavelength of 10^{-6} m (or 0.01 mm). What type of electromagnetic radiation is it?

4 Which radiations have an ionising effect?

★ Stretch Yourself

Microwaves from Source A have a frequency of 3 GHz and from Source B have a frequency of 30 GHz.

1 Which microwaves have the most energy?

2 Calculate the wavelength of the microwaves from A and B.

Light, Radio Waves and Microwaves

Visible Light

When electromagnetic radiation from the Sun arrives at the top of the atmosphere some of it is reflected, some is transmitted to the Earth's surface and some is absorbed by the atmosphere.

The electromagnetic radiation that arrives at the Earth's surface is:
- High energy infrared.
- Visible light.
- Low energy ultraviolet.

Some of the ways visible light is used include:
- By our eyes when images are formed on the retina at the back of the eye.
- Photography when images are formed on the light sensitive film in a camera.
- Digital photography when images are formed on the light sensitive screen and stored electronically.

- Plants use light as an energy source for photosynthesis.

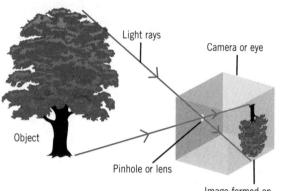

How an image is formed

Light rays

Camera or eye

Object

Pinhole or lens

Image formed on light sensitive surface

✓ Maximise Your Marks

Light travels in straight lines so always draw rays of light with a ruler.

Build Your Understanding

The eye is similar to a camera. In both, light rays pass through a small hole and produce a small upside down image. A lens can be used to gather and focus more light.

The image is formed on a light sensitive surface:
- On photographic film a chemical reaction occurs which changes the colour of the film.

- On the cells of the retina a chemical reaction causes electric signals to travel along the optic nerve to the brain.
- On the screen inside a digital camera electronic signals are sent to the memory. The amount of information to store the picture is measured in bytes.

Radio Waves

Radio waves are produced when an alternating current flows in an aerial. They spread out and travel through the atmosphere.

Another aerial is used as a detector. The radio waves produce an alternating current in it, with a frequency that matches that of the waves.

Transmitting and receiving radio waves

Changing currents in the transmitter

Radio waves

Changing currents in the aerial

Transmitter

Radio receiver

Radio Waves and Microwaves

Some important properties of radio waves and microwaves are:

- They are reflected by metal surfaces.
- They heat materials if they can make particles in the material vibrate.
- The amount of heating depends on the power of the radiation and the time that the material is exposed to the radiation.

Microwaves

Microwaves are transmitted through glass and plastics. They are absorbed by water, though how well depends on the frequency (energy) of the microwaves.

Microwave ovens use a microwave frequency which is strongly absorbed by water molecules, causing them to vibrate, increasing their **kinetic energy**. This heats materials containing water, for example food. The microwaves penetrate about 1 cm into the food. Conduction and convection processes spread the heat through the food.

Microwave oven radiation will heat up our body cells and is very dangerous at high intensity because it will burn body tissue. The radiation is kept inside the oven by the reflecting metal case and metal grid in the door.

How a microwave oven works

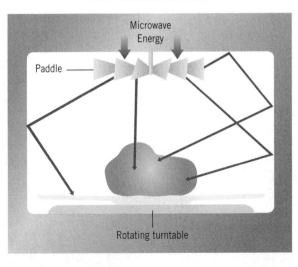

Microwave Energy

Paddle

Rotating turntable

Build Your Understanding

Radio waves are transmitted through the atmosphere without being absorbed. Medium wavelength radio waves are reflected from the ionosphere, which is a layer of charged particles in the upper atmosphere.

Most microwaves are transmitted through the atmosphere, but some wavelengths are absorbed or scattered by dust and water vapour in the atmosphere and also by water droplets in rain and clouds. They pass through the ionosphere without being reflected.

How radio waves and microwaves travel

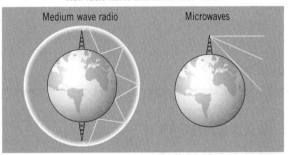

Medium wave radio   Microwaves

💡 Boost Your Memory

Make a table of materials that reflect, absorb and transmit light, radio waves and microwaves. This will help you to remember the similarities and the differences.

❓ Test Yourself

1. Give two differences between an object you look at and the image formed on your retina.

2. What material would you use for a cooking container for use in a microwave oven? Explain your choice.

3. An 850 W microwave oven heats food for 10 seconds. How much energy in joules heats the food?

4. Why are the microwaves used in ovens especially dangerous to the human body?

⭐ Stretch Yourself

1. How could microwaves that are absorbed by water droplets be used in weather forecasting?

Wireless Communications 1

Electromagnetic Waves

Radio Waves

Radio waves are used for **broadcasting** radio and TV programmes. Anyone with a receiver can tune it to the radio frequency to pick up the signal.

When radio stations use similar transmission frequencies the waves sometimes **interfere** with each other. Medium wavelength radio waves are reflected from the ionosphere so they can be used for long distance communication, but not for communicating with satellites above the ionosphere.

Microwaves

The transmitter and receiver must be in line of sight, (one can be seen from the other), so they are positioned high up, often on tall masts. They must be close together so that hills, or the curvature of the Earth, cannot block the beam.

Signals are sent to and from **satellites,** which relay signals around the Earth. This may be for TV programmes, telephone conversations or monitoring the Earth, for example weather forecasting.

Satellite communication

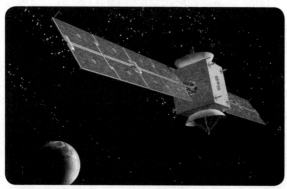

✔ Maximise Your Marks

Make sure you can compare the use of microwaves and radio waves. In 'compare' answers refer to both, for example microwaves pass through the ionosphere, but medium wave radio waves do not.

Diffraction

Diffraction is the spreading of a beam through gaps and around corners (see page 21). The maximum effect occurs when the gap has a similar size to the wavelength.

Radio waves of about 5 m are diffracted by large buildings. Radio waves of 1 km are be diffracted around hills and through valleys, so they are able to reach most areas and are suitable for broadcasting.

Microwave beams of a few centimetres do not spread round corners or around hills. This is why the transmitters and receivers must be in line of sight. When microwaves are transmitted from a satellite dish the wavelength must be small compared to the dish diameter to reduce diffraction. This means that, compared to radio waves, microwaves can be sent as a thin beam.

Diffraction of microwaves and radiowaves

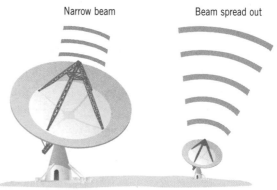

Narrow beam   Beam spread out

Large dish   Small dish

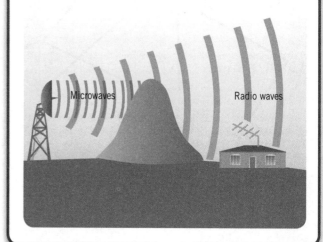

Microwaves   Radio waves

Analogue and Digital Signals

Radio waves, microwaves, infrared and visible light are all used to carry information, which can be sound, pictures or other data. The information is called the **signal**. It is added to an electromagnetic wave called the **carrier wave** so that it can be transmitted. When the wave is received the carrier wave is removed and the signal is reconstructed. There are two types of signal, **analogue** and **digital**.

An analogue signal changes in frequency and/or amplitude continually in a way that matches changes in the voice or music being transmitted.

A digital signal has just two values – represented as 0 and 1 (or on and off). The signal is converted into a code of 0s and 1s. It becomes a stream of 0 and 1 values. These pulses are added to the carrier wave and transmitted. After the signal is received it is decoded to recover the original signal.

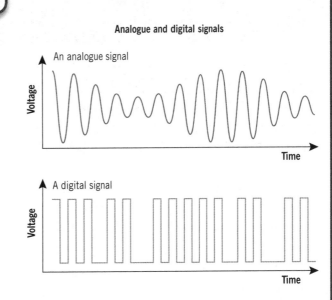

Analogue and digital signals

An analogue signal

A digital signal

✓ Maximise Your Marks

A common mistake is to think that we can hear radio waves. We cannot hear any electromagnetic radiation. The radio waves are the carrier used to carry a signal that is converted into a sound wave by the receiver.

Build Your Understanding

Both analogue and digital signals can pick up unwanted signals that distort the original signal. These unwanted signals are called noise.

Digital signals give a better quality reception because the noise is more easily removed.

They can be cleaned up in a process known as regeneration. The 0 or 1 values can be restored because each pulse must be a 0 or a 1. Analogue signals can be amplified, but the noise is amplified too.

❓ Test Yourself

1. What type of electromagnetic waves are used to **a)** broadcast TV and **b)** for satellite TV?

2. Why are there often lots of microwave transmitters on the top of hills?

3. What is the difference between an analogue signal and a digital signal?

4. In communication signals, what is noise?

⭐ Stretch Yourself

1. Why are satellite dishes larger than the wavelength of microwaves they transmit?

2. Television pictures from analogue signals often have white speckles on the picture, but digital pictures do not.
   a) What is this effect called?
   b) Why does it only happen to analogue pictures?

Wireless Communications 2

Wireless and Mobile Phones

Wireless communication uses microwaves and radio waves to transmit information. The advantages of this are:

- No wires are needed to connect laptops to the internet, or for mobile phones or radio.
- Phone calls and e-mail are available 24 hours a day.
- Communication with wireless technology is portable and convenient.

Mobile phones use microwave signals. The signals from the transmitting phones are reflected by metal surfaces and walls, and travel through the air to communicate with the nearest transmitter mast.

There is a network of transmitter masts to relay the signals on to the nearest mast to the receiving phone.

It is unclear whether there are any long-term effects of using mobile phones. There may also be a risk to residents living close to mobile phone masts.

Most people consider the risks and benefits, and decide that the benefit of using a mobile phone outweighs the risk.

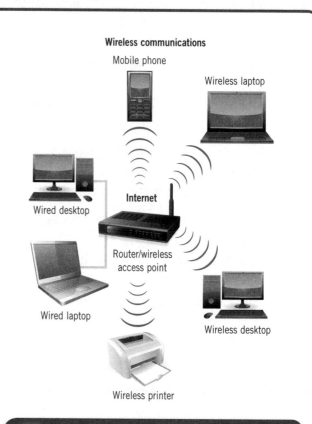

Wireless communications

Mobile phone

Wireless laptop

Wired desktop

Internet

Router/wireless access point

Wired laptop

Wireless desktop

Wireless printer

✓ Maximise Your Marks

Remember that the danger of microwaves and other electromagnetic radiation at the low energy end of the spectrum depends on its intensity. Do not confuse this with the danger of the ionising radiations at the high energy end of the spectrum.

Build Your Understanding

We have only limited data about the possible dangers of mobile phones. The transmitter is held close to the user's head, so the microwaves must have a small heating effect on the brain. There is no evidence that this is dangerous. But if there is a long-term risk, there has not yet been enough time to collect the evidence. We need to collect and analyse data for an average lifetime of about 80 years before we can say whether there is any evidence of a long-term risk.

The Health Protection Agency (HPA) is made up of independent scientists, who look at the

evidence of the effects of radiation on health. So far studies have not found that mobile phone users have suffered any serious ill effects. Their advice is to limit the use of handsets held close to the head, especially for young children in case there are long-term effects.

Scientists do lots of studies to see if the results are repeatable. We can have more confidence in a study if the results been replicated by the scientists themselves and reproduced by other scientists.

Correlation Between Factors

A good study uses a **large sample** of thousands of people. The sample will be matched, so that scientists compare the same type of people who use mobile phones with people who don't, to see if this is a **factor** that increases the risk. For example, they might compare women, children, or men who smoke. If they find a **correlation**, they will look to see if this is due to a different **cause**. A correlation is a link between two factors.

If there is a correlation it does not always mean that one factor causes the other. For example, there is a correlation between increased sales of ice cream and an increase in hay fever, but this does not mean that eating ice cream causes hay fever. Both increase in hot sunny weather.

It is important to scientists to find a **mechanism** that explains why a factor causes an **outcome**. For example, the mechanism by which radioactive materials cause cancer is the radiation ionising atoms and damaging body cells.

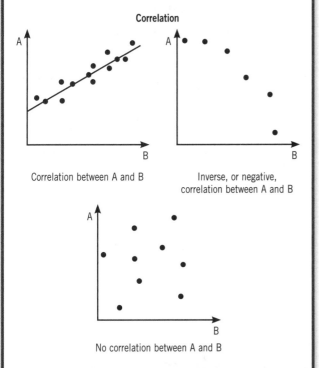

Correlation

Correlation between A and B

Inverse, or negative, correlation between A and B

No correlation between A and B

⚲ Boost Your Memory

Remember the ice cream example so that you can use it to explain why correlation is not the same as cause.

Build Your Understanding

It is very difficult to control other factors in a study of people using mobile phones because there are so many things people do differently. Laboratory investigations should be designed to control all the factors.

Example: In an experiment to see how the distance microwaves travel in glass affects their absorbtion, you would make sure that these factors are kept constant:

- The intensity of microwaves from the source.
- The detector.
- The distance the microwaves travel in air before and after the glass.
- The type of glass.

❓ Test Yourself

1 Give one advantage of wireless technology.

2 Give an example of a benefit and a risk of using a mobile phone.

3 Sketch a graph showing how ice cream sales and hay fever might be related.

4 What affect do microwaves have that might be a mechanism for causing damage to the body?

★ Stretch Yourself

1 Why do scientists think that young children should use hand held mobile phones as little as possible?

2 In the example of the experiment to see how distance affects absorption of microwaves:

   a) What is the factor and what is the outcome?

   b) Describe a correlation you might expect to find.

   c) What would be the effect of **not** controlling each factor?

Infrared

Uses of Infrared

Our skin detects **infrared** radiation and we feel **heat**.

All objects emit electromagnetic radiation that depends on their temperature. Hot objects glow red and very hot objects glow white hot, because they emit light as well as infrared. The hotter the object, the more electromagnetic radiation it emits and the higher the maximum frequencies.

Warm objects, like radiators and human bodies, emit infrared. Night-vision goggles, cameras and thermograms use false colour so that we can 'see' the radiation.

A thermogram of an elephant

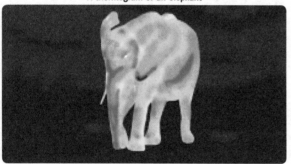

Infrared emission, absorption and reflection depends on the surface (see page 14). When infrared is absorbed the particles in the surface vibrate and gain kinetic energy. This is why infrared radiation is used for cooking. The surface of food gets hot and then the inside is heated by convection and conduction. We must be careful that intense infrared radiation does not have this effect on our skin and cause burning. Some cooking appliances, like grills and toasters, emit red light as well as infrared.

Infrared radiation is used in **remote controls** for televisions and other electronic appliances. If you look at the beam from a remote control through a digital camera (which shows up infrared signals that our eyes cannot see) you can see the flashing infrared emitting diode sending the **digital signal**. These signals cannot pass through solid objects, but sometimes reflect off walls and ceilings to operate the television.

Build Your Understanding

Infrared sensors detect human body heat, so they are used in security systems. An infrared beam can be used as part of a burglar alarm. The burglar cannot see the beam and steps into it. This blocks the radiation reaching the sensor and triggers the alarm.

Night-vision goggles and cameras work by detecting the infrared radiation emitted by objects at different temperatures. The different intensities of infrared are changed to different intensities of visible light, or often to different colours to help us see the objects even more clearly. Warm objects, like humans, other animals and cars, can be easily picked out.

✓ Maximise Your Marks

Remember that infrared comes between microwaves and red light. Make sure you can explain the differences between them, especially their properties and uses.

Communications

Infrared radiation and light are both transmitted along glass optical fibres by being reflected as shown in the diagram.

Light travels along optical fibres

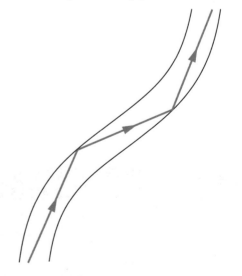

Optical Fibres

The fibres are made with a core that has a different refractive index to the outer cladding. Total internal reflection occurs at the boundary between the materials.

Digital signals can be sent for long distances using **optical fibres** because glass is transparent to visible light and infrared. When the signal needs boosting, digital signals are easily regenerated. A stream of data can be transmitted very quickly. Both infrared and red light **lasers** are used as radiation sources for fibre optic communications. Lasers produce a very narrow beam of intense radiation of one colour.

Total internal reflection can happen when light is refracted at a boundary with a less dense medium.

When the angle of incidence equals the **critical angle** the angle of refraction is 90°.

When the angle of incidence is greater than the critical angle total internal reflection occurs.

Optical fibres

Build Your Understanding

Advantages of fibre optic communication include:

- Infrared signals in fibre optic cables experience less interference (they pick up less noise) than microwaves passing through the atmosphere.
- Digital signals can be multiplexed, which is when signals are divided into sections and sent alternately so that many different signals can be sent along one fibre at the same time.

Lasers produce a beam with low divergence (does not spread out) in which all the waves have the same frequency and are in phase (in step) with each other.

Lasers are also used in a compact disc (CD) player. CDs store information digitally as a series of pits and bumps (0s and 1s) on the shiny surface.

A laser beam is reflected differently from the pits and bumps and a detector is used to 'read' the different reflections, reproducing the 0s and 1s of the signal.

💡 Boost Your Memory

Draw a diagram of a signal from a mobile phone travelling through a fibre optic link and a microwave link that includes a satellite. Label the diagram and it will help you to revise all the parts of microwave communication.

❓ Test Yourself

1. How do we detect infrared radiation?
2. How does a toaster cook a slice of bread?
3. Explain why the elephant's ear in the picture on page 34 is blue while the rest of its body is orange/red.
4. Explain how the signal from a TV remote control would be different for selecting two different channels.

⭐ Stretch Yourself

1. How is light from a laser different to light from a light bulb?
2. Give an advantage of using infrared radiation and fibre optic cable over using a microwave link.

The Ionising Radiations

Electromagnetic Waves

Ionising Radiation

Gamma rays, **X-rays** and high energy **ultraviolet radiation** are high energy radiations which can **ionise** atoms they hit. Atoms are ionised when electrons are removed and this makes them charged and more likely to take part in chemical reactions.

If these atoms are ones inside cells of the body, ionising radiation can damage or kill them. If the DNA in a cell is damaged it may mutate. This can cause cells to grow out of control which means they have become cancer cells.

Radioactive Emissions

Some materials are **radioactive.** This means they randomly emit ionising radiation from the nucleus of unstable atoms. There are three types:
- Gamma rays which are electromagnetic waves.
- **Alpha particles** which are helium nuclei. They are positively charged.
- **Beta particles** which are high energy electrons from the nucleus. They are negatively charged.

There are two types of danger from all these radioactive materials:
- **Irradiation** is being exposed to radiation from a source outside the body.

- **Contamination** is swallowing, breathing in, or getting radioactive material on your skin.

A short period of irradiation is not as dangerous as being contaminated because, once contaminated, a person is continually being irradiated.

Uses of ionising radiation:
- Gamma rays are used for sterilising medical equipment and killing cancer cells.
- Alpha emitters are used in smoke detectors.
- Beta emitters are used as tracers.

X-Rays

X-rays have high energy and pass through the body tissues. They are stopped by denser materials such as the bones, and pieces of metal. They are used to image the body.

The figure shows X-rays directed towards the patient with a photographic plate placed behind. The plate darkens where the X-rays strike it and there are white shadows where the bones absorbed the X-rays.

X-rays are also used for security scans of passengers' luggage. Metal items and batteries block the X-rays and show up as shadows on the screen.

There are benefits to using ionising radiation but these must be weighed against the risks. To protect people dense materials, like lead screens

and concrete, are used to absorb the radiation and act as barriers. Radiation workers, like radiographers and workers at nuclear power stations, are monitored to make sure they are not exposed to high levels of radiation. They wear protective clothes.

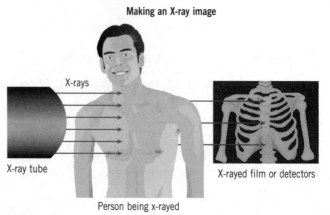

Making an X-ray image

X-rays

X-ray tube

X-rayed film or detectors

Person being x-rayed

Benefits and Risks

A patient should not receive too many X-rays. They are used when the benefit (for example, finding a broken bone) is greater than the risk.

In the 1950s all pregnant women were X-rayed to check on the development and position of the baby. This caused a few cancers among children. The benefits were not greater than the risks and routine X-rays were stopped.

Radiation workers do not benefit from a dose of radiation and would be at risk from high exposure every day. Radiographers leave the room, or wear a lead apron, when a patient is X-rayed.

People tend to overestimate the risks from unfamiliar things compared to familiar things (for example they may think flying is more risky than cycling.) Because ionising radiation can't be seen, people tend to think it is more dangerous than it is.

✓ Maximise Your Marks

Whether radiation is dangerous sometimes depends on whether a person is contaminated or irradiated.

Make sure you can explain the difference.

Ultraviolet Radiation

The lower energy ultraviolet radiation that reaches the Earth's surface can cause:
- premature skin aging
- suntans
- sunburn
- skin cancer
- damage to the eyes.

The number of cases of skin cancer has increased in recent years as people spend more time in the sun. Dark skins absorb more ultraviolet radiation than light skins, which are more easily damaged.

To reduce the risk people should stay out of the sun during the hottest part of the day and cover up with a hat and clothes. They should use a high protection factor sunscreen. Ultraviolet blocking sunglasses are recommended to protect the lens of the eye from damage.

Ultraviolet radiation is used to detect forged bank notes because genuine bank notes have some features that **fluoresce** in ultraviolet radiation.

Build Your Understanding

There are benefits of spending time in the sun, for example our skin makes vitamin D which reduces the risk of other cancers, so benefits must be weighed against risks.

✓ Maximise Your Marks

In a question about advantages and disadvantages, or about benefits and risks, you must give both to get full marks. A list of benefits cannot be awarded a mark for stating a risk.

? Test Yourself

1. What does it mean if radiation is ionising?
2. What is the difference between being contaminated or irradiated by radioactive material?
3. Why are babies no longer routinely X-rayed in the womb?
4. Which type of electromagnetic radiation tans the skin?

★ Stretch Yourself

1. How does leaving the room when an X-ray is taken protect a radiographer?
2. Give a benefit and a risk of sunbathing.

The Atmosphere

Absorption of Radiation

The Earth is surrounded by an atmosphere made up of different gases. It allows some frequencies of electromagnetic radiation from the Sun to pass through (see page 28).

The Earth emits infrared radiation at lower frequencies. These frequencies are absorbed by some gases in the atmosphere such as **carbon dioxide**, **water vapour** and **methane**. This keeps the Earth warm and is called the atmospheric **greenhouse effect**. Without it the Earth would be much colder – too cold for some species to survive.

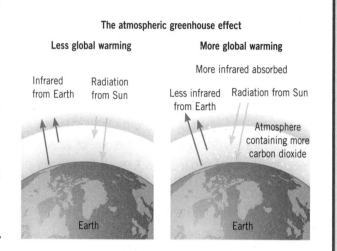

The atmospheric greenhouse effect

Less global warming

Infrared from Earth   Radiation from Sun

Earth

More global warming

More infrared absorbed

Less infrared from Earth   Radiation from Sun

Atmosphere containing more carbon dioxide

Earth

Build Your Understanding

High energy infrared and visible radiation from the Sun passes through the glass into a greenhouse. Low energy infrared radiation from inside the greenhouse cannot escape through the glass. This keeps the greenhouse warm and is called the greenhouse effect.

Global Warming

Scientists agree that **global warming** is happening. The Earth is getting warmer.

In the last 200 years the amount of carbon dioxide in the atmosphere has steadily increased.

Reasons include:
- Burning **fossil fuels**.
- Clearing forests so that fewer trees are using carbon dioxide for **photosynthesis**.

Computer climate models show that human activities are increasing greenhouse gases and causing global warming, but some scientists do not agree. They do not agree about how much change is likely and what effect it will have. It could result in:
- Extreme weather conditions in some regions.
- Rising sea levels due to melting ice and expansion of water in oceans which may flood low lying land.
- Some regions no longer able to grow food crops.

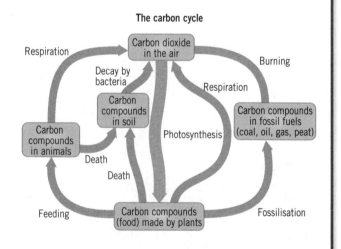

The carbon cycle

Respiration

Carbon dioxide in the air

Burning

Decay by bacteria

Respiration

Carbon compounds in soil

Carbon compounds in animals

Carbon compounds in fossil fuels (coal, oil, gas, peat)

Photosynthesis

Death

Death

Feeding

Carbon compounds (food) made by plants

Fossilisation

✓ Maximise Your Marks

A common mistake is to muddle up the effect of the ozone layer on ultraviolet and the effect of infrared on the greenhouse effect. Make sure you understand the difference.

Ultraviolet in the Atmosphere

Some ultraviolet radiation from the Sun reaches the surface of the Earth but, fortunately for living things, the highest energy (highest frequency) ultraviolet radiation is stopped by the **ozone layer**.

This is a layer of ozone gas in the upper atmosphere that absorbs ultraviolet radiation and chemical changes occur. Without this layer higher frequency ultraviolet would cause more sunburn, skin cancer and cataracts.

In 1985 an ozone hole was discovered above the Antarctic. Scientists had been looking for a reduction in the ozone layer, but did not expect to find a large hole that formed so quickly.

The measurements were repeated with new equipment. They were replicated by the scientists who discovered the hole and reproduced by other scientists. Where the ozone layer has been depleted, living organisms, especially animals, suffer more harmful effects from ultraviolet radiation.

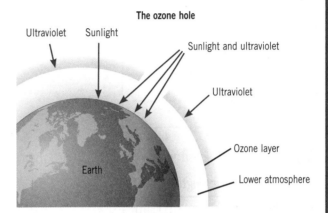

The ozone hole

Ultraviolet | Sunlight | Sunlight and ultraviolet | Ultraviolet | Ozone layer | Lower atmosphere | Earth

Build Your Understanding

The explanation that scientists had for the ozone hole was that using gases called CFC gases in aerosol cans (as the propellant to force the contents of the can out of the nozzle) and as the refrigerant in refrigerators and freezers, caused the concentration of CFC gas in the atmosphere to increase. This pollution reacted with the ozone in the springtime and reduced the amount in the ozone layer. Scientists predicted that the ozone hole would grow. This prediction was correct.

International agreements, like the Montreal Protocol in 1987, have stopped the use of CFCs and other ozone depleting gases. This is a good example of the world's governments working together. The ban is having an effect and the hole is getting smaller, but the CFCs are very stable so it will be at least 50 years before the hole disappears.

? Test Yourself

1. Which gases in the atmosphere contribute to global warming?

2. Why is there more carbon dioxide in the atmosphere today than there was 100 years ago?

3. Which gas absorbs ultraviolet radiation and protects living things?

★ Stretch Yourself

1. Most scientists agree that the climate is changing. What do some disagree about?

2. Which gases cause the hole in the ozone layer and what has been done about this problem?

Practice Questions

Complete these exam-style questions to test your understanding. Check your answers on page 122. You may wish to answer these questions on a separate piece of paper.

1 Draw lines to join the type of electromagnetic waves to the use. (1)

Type
Microwaves
Radio waves
Infrared
X-rays

Use
Communication with satellites
Optical fibre communications
Medical scans
TV broadcasts

2 Draw lines to join the wavelength to the type of electromagnetic waves. (1)

Wavelength
500 nm (1nm = 10^{-9}m)
0.01 mm
3 cm
300 m

Type
Infrared
Microwaves
Radio waves
Visible light

3 If 100 J of radiation illuminates one square metre of a surface every second, what is the intensity of the light? (2)

4 Explain why the residents of a village in a valley surrounded by hills can receive satellite TV, but not terrestrial TV broadcasts. (2)

5 Explain the difference between an analogue and digital signal. (2)

6 Scientists set up a study to see if brain tumours in children are increased by using mobile phones. Here are three possible samples for a ten-year study:

X 100 children aged between 8 and 12 – 50 use a mobile phone, while 50 do not.

Y 1000 children aged between 0 and 15 – 500 use a mobile phone at the start of the trial, by the end of the trial 950 use one.

Z 1000 children aged between 8 and 12 – 400 use a mobile phone at the start, by the end of the trial 600 use one at.

a) Give an advantage and disadvantage of sample X. (2)

b) Give an advantage and disadvantage of sample Y. (2)

c) Explain how sample Z could be used to produce a better sample of 800 for comparison. (2)

d) Explain whether you think 10 years is a reasonable time for the study. (2)

e) Some children use a mobile phone for several hours of conversation a day, others only use it to text. Explain how this will affect the trial. (2)

7 Explain two ways that a TV signal can be sent to and received by homes in the UK.
(In your answer, aim to write about 8 and 12 lines.) (6)

How well did you do?

| 0–7 | Try again | 8–14 | Getting there | 15–19 | Good work | 20–24 | Excellent! |

The Solar System

The Solar System

The **Solar System** was formed over a very long time from **clouds** of **gases and dust** in space, about **five thousand million years ago**.

The **planets** orbit the **Sun**, which is the star at the centre of the Solar System. There are eight planets and Pluto which is a dwarf planet. Some planets are orbited by one or more **moons**. An object which orbits another is called a **satellite**. The **Moon** is a natural satellite of the **Earth**.

Asteroids are rocks, up to about 1 km in diameter, that orbit the Sun. These have been around since the formation of the Solar System. Most of these are between Mars and Jupiter. Jupiter is the largest planet and there is a large gravitational force towards it. This has prevented the formation of a planet between Mars and Jupiter, in the asteroid belt.

There are many **comets**. Some take less than a hundred years to orbit the Sun, while others take millions of years. They are made of ice and dust. Most have a nucleus of less than about 10 km, but this vapourises and becomes a cloud thousands of miles across when the comet is close to the Sun. Comets spend most of their time far from the Sun – much further away than the dwarf planet Pluto.

Meteors, or shooting stars, are caused by dust and small rocks, usually from a comet. When the Earth passes through this debris the rocks fall through the atmosphere. They heat up and glow. Any pieces that land on the Earth are called **meteorites**.

Near Earth Objects (NEOs) are comets and asteroids in an orbit that brings them close to the Earth. Some of them could one day collide with Earth. The **craters** on the Moon are evidence of collisions in the past. Craters on the Earth have been mostly eroded away.

Surveys by **telescopes** try to observe and record the paths (trajectories) of all NEOs. They can be monitored from Earth or by satellite. We can make sure that we have advance warning of a collision. The idea of deflecting a NEO using explosions is being considered. At present scientists are collecting information about the composition and structure of NEOs. The possibility of destroying one is still only in the planning stage.

The planets in the Solar System

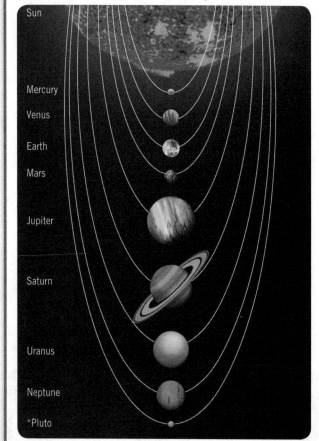

Sun
Mercury
Venus
Earth
Mars
Jupiter
Saturn
Uranus
Neptune
*Pluto

💡 Boost Your Memory

To remember the order of the planets and Pluto (Mercury, Venus, Earth, Mars, Jupiter, Saturn, Uranus, Neptune and Pluto) use a mnemonic like:

My Very Easy Method Just Speeds Up Naming Planets

✓ Maximise Your Marks

Do not confuse asteroids, meteors, comets and meteorites.

Gravity and Orbits

The **force of gravity** is an attractive force between the Sun and a planet. It keeps the planet moving in **orbit** around the Sun. For an object to move in a circle there must be a force on it towards the centre of the circle. This force is called the **centripetal force**.

If an object moves closer to the Sun the gravitational force on it will increase and it will speed up.

Comets have very **elliptical orbits**. They are kept in orbit by the gravitational force of attraction to the Sun, but their distance from the Sun changes as shown in the diagram. The force on the comet is largest close to the Sun, where the distance is smallest. The speed of the comet is much greater close to the Sun.

A comet's orbit

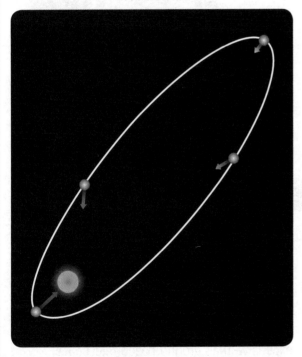

Models of the Solar System

In Ancient Greece scientists thought that the Earth was stationary and that the Sun, Moon and planets moved around the Earth. This is known as the **geocentric** model. As scientists made more accurate and detailed observations of the Solar System they noticed that the planets moved in complicated patterns. A simpler explanation was that all the planets, including Earth, orbited the Sun. This is called the **heliocentric** model.

Build Your Understanding

The geocentric model of the Earth at the centre of the Universe developed by Ptolemy is also called the Ptolemaic model. Copernicus was the first person to write down a suggested heliocentric model. Galileo used a new telescope to observe moons orbiting Jupiter which added to the evidence that everything did not have to orbit the Earth. This new evidence was not accepted for many years because it did not agree with religious beliefs at the time.

The outer planets

❓ Test Yourself

1. Which is the largest planet?
2. What is the difference between an asteroid and a comet?
3. Describe the heliocentric model of the Solar System.
4. What is the nearest star to Earth?

⭐ Stretch Yourself

1. What force keeps a comet moving around the Sun?
2. What are the main differences between the orbit of a comet and the orbit of the Earth?

Space Exploration

Exploring the Solar System

Methods of finding out about the Solar System and what is beyond it are:

- To use telescopes.
- To send unmanned space probes.
- To send manned spacecraft.

Beyond the Solar System using a telescope is the only way to study the Universe, because of the huge distances and time needed to travel to them.

Telescopes are also useful for study of the Solar System. They can be positioned on Earth where they are easier to maintain and repair, but where they must look through the **atmosphere**.

Some telescopes are positioned high up on mountain tops to get above the clouds, dust and pollution, and away from city lights.

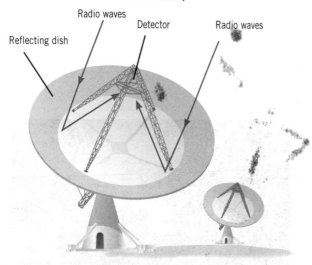

A radio telescope

Reflecting dish
Radio waves
Detector
Radio waves

Telescopes in orbit avoid these problems and have a much clearer view of the stars. Launching a telescope into orbit is more expensive than building one on Earth.

There are telescopes designed to use all types of **electromagnetic waves**, including **radio telescopes** which observe radio waves from space. As X-rays do not pass through the atmosphere, **X-ray telescopes** are always placed in orbit.

Hubble space telescope

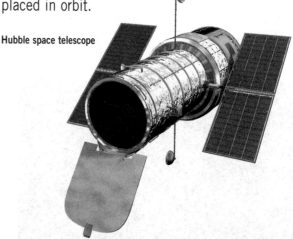

✓ Maximise Your Marks

Remember if you are asked to compare telescopes on Earth and in orbit, or manned and unmanned space missions, that you must compare the two. Saying that telescopes in orbit are expensive will gain no marks. You must say that telescopes in orbit are more expensive than ground based telescopes.

Unmanned Missions

Many unmanned space probes have been launched. NASA spacecraft and Mars rovers have studied Mars. We have lots of information about Mars without humans travelling there.

Unmanned spacecraft can operate unharmed in a lot of conditions that would kill humans. By using remote sensors and computers they can send back information on:

- Temperature, radiation, gravity and magnetic field.
- Gases in the atmosphere.
- Composition of the rocks.
- Appearance of the surroundings (using TV cameras).

Manned Missions

Spacecraft have put men and women in orbit and on the Moon. There are plans to send humans to Mars, but the difficulties of space missions to other planets are extreme. These include:

- Enough food, water, oxygen and fuel must be carried for the entire trip, including delays.
- The distance means that emergency supplies could not be sent in time to be of use.
- An artificial atmosphere must be set up inside the spacecraft, and levels of carbon dioxide and oxygen monitored.

- Interplanetary space is very cold. There must be heating to keep astronauts warm. In direct radiation from the Sun the spacecraft will heat up, so cooling is required.
- There will be very low gravity during the journey. Bones lose density and muscles waste away under these conditions. A special exercise programme is needed using exercise machines.
- Astronauts must be shielded from cosmic rays and from the radiation released from the Sun when there is a solar flare.

Build Your Understanding

Is it best to send unmanned space probes or a manned mission to find out about Mars?

Points to consider are:

- The costs of unmanned missions are less.
- There are no lives at risk if something goes wrong with an unmanned mission.
- Everything must be planned before an unmanned mission leaves – no adjustments,

repairs or changes can be made unless they are programmed into the computers and can be done remotely.

- Most people are more interested in manned missions. Unmanned missions do not inspire people in the way that manned missions do.

The Moon

The current theory of how the Moon was formed is that the Earth collided with another planet, about the size of Mars.

Most of the heavier material of the other planet fell to Earth after the collision and the iron core of the Earth and the other planet merged. Some less dense material was thrown into orbit and formed the Moon.

Evidence for this theory is:

- Samples of Moon rocks brought back to Earth by astronauts show that Moon rocks are the same as Earth rocks – unlike rocks from other planets and moons.
- The Moon is completely made of less dense rocks – there is no iron core – unlike other planets, moons and asteroids.
- The Moon has no recent volcanic activity, but its rocks are igneous.

? Test Yourself

1. List two ways that scientists have collected information about Mars.

2. Telescopes can be Earth based or launched into space. Give one problem using:

   a) An Earth based telescope.

   b) A space based telescope.

3. How do scientists think the Moon formed?

★ Stretch Yourself

1. Give an advantage and a disadvantage of manned space missions compared to unmanned space missions.

2. What evidence do we have that the Moon formed during the collision of a planet with Earth?

A Sense of Scale

The Light Year

A **light year** is the distance light travels in a year.

Light travels through space, which is a vacuum, at 300 000 km/s. Light from the Sun arrives on Earth after about eight minutes. After a year, light from the Sun will have travelled a distance of:
= 300 000 km/s × (the number of seconds in a year)

This distance is about 9.5×10^{12} m – about nine and a half million million kilometres. (You don't need to remember this number.)

Distances to stars and galaxies are so large that we measure them in light years. The nearest star to Earth is the Sun. The second nearest is about four light years from Earth.

☉ Boost Your Memory

To remember the increasing size; A, (no B), C, D, E. (**a**steroid, **c**omet, **d**warf Pluto, **E**arth).

Increasing size ↓	Object	Size or distance	Approximate age
	Asteroids and meteors	Usually less than 1 km diameter	
	Comets	Usually less than 15 km diameter	
	Dwarf planet, Pluto		
	The Moon	Smaller than Earth (diameter is about a quarter of the Earth's)	4500 million years
	Earth	Diameter 12 760 km	4500 million years
	Largest planet, Jupiter	Diameter over 10 times the Earth's	
	Sun	Diameter over 100 times the Earth's	5000 million years
	Earth's orbit		
	Solar System	Diameter about 10 000 000 000 km Diameter about 10 light years	4600 million years
	Next nearest star (after the Sun)	About four light years	
	Milky Way galaxy	Diameter about 100 000 light years	
	Milky Way to nearby galaxies	About 100 000 light years	
	The observable Universe	About 14 thousand million light years	About 14 thousand million years

Looking Back in Time

When you look through a telescope and see a star 100 light years away, what you see – the light entering your telescope – left the star 100 years ago. So looking at very distant planets is like looking back in time. Distant objects look younger than they really are.

If an alien 950 light years away could look at Earth through its telescope on the right day in 2016 it could watch the Battle of Hastings in 1066.

✓ Maximise Your Marks

The light year is a unit of distance, not a unit of time.

Build Your Understanding

The parallax method uses the fact that the Earth orbits the Sun. One observation of the night sky is made and then a second six months later. The Earth has changed its position in space by a distance equal to the diameter of its orbit. However, most stars are so far away that the distance moved is too small to measure.

The difficulties in making observations lead to uncertainty in our measurements of the distances to stars and galaxies. Another difficulty for astronomers is the amount of light pollution from the Earth. All the light from our cities shines into the night sky and makes it difficult to pick up the very weak signals from distant stars and galaxies.

The Parallax Method

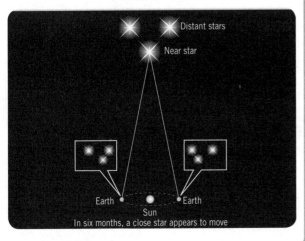

Distant stars

Near star

Earth   Earth
Sun
In six months, a close star appears to move

Brightness

A very bright star may look bright because:
- It is bigger than other stars.
- It is hotter than other stars.
- It is closer than other stars.

If there are reasons to believe stars are similar (for example, they are the same colour) then a difference in **brightness** can be used to measure the distance. The further away the star the dimmer it is.

Parallax

You can see the **parallax** effect from a train or car window; objects close to you seem to move more quickly and change position, when compared to distant objects. When two observations of the night sky are taken six months apart, so that the Earth has moved, a star that is close to Earth will have changed its position when compared to distant stars in the background. The amount the nearby star has moved is used to calculate how far it is from Earth.

? Test Yourself

1 List the following objects in order of increasing size: a comet, the Universe, the Earth, the Moon, the Milky Way, the Sun.

2 How long does it take light to cross the Solar System?

3 A student notices that looking out of two different classroom windows a lamp post appears to move against the background more than a bus stop sign. Which is closer?

★ Stretch Yourself

1 Explain one problem with using the brightness of a star to work out its distance from Earth.

2 Explain why the parallax method can only be used for nearby stars.

Stars

The 'Lifetime' of a Star

Formation:

- Stars begin as large clouds of dust, hydrogen and helium called an **interstellar gas cloud** or a **nebula**.
- Gravity makes the nebula contract, this makes it heat up and it is now a **protostar**.
- When it is hot enough **hydrogen nuclei fuse** together to form **helium nuclei**. This is called **nuclear fusion**. Energy is released as light and other electromagnetic radiation. A **star** has formed.

Stable lifetime, fusing hydrogen:

- The star is one of a large number of **main sequence stars** like our Sun. Stars spend a long time fusing hydrogen. Our Sun will do this for ten thousand million years.
- What happens when a star has fused most of its hydrogen depends on the mass of the star.

Final stages of a small star like our Sun:

- The star cools, becoming redder and expands to form a **red giant**. The core contracts and helium nuclei fuse to form carbon, oxygen and nitrogen.

- After all the helium has fused the star contracts and the outer layers are lost. As these outer layers move away they look to us like a disc that we call a **planetary nebula**.
- The remaining core becomes a small, dense, very hot **white dwarf**. The star will then cool over a very long time and become a **brown or black dwarf**.

Final stages of a massive star:

- The star cools and expands to form a **red supergiant**. The core contracts and **fusion** in the core forms carbon and then heavier elements up to the mass of iron.
- When the nuclear fusion reactions are finished the star cools and contracts, which heats it again until it explodes. The explosion is called a **supernova** and is the largest explosion in the Universe. All the elements heavier than iron that exist naturally on planets were created in supernova explosions.
- The core is left as a **neutron star**. It is very dense. It has a large mass, but a very small volume. If the star is very massive, the core is left as a **black hole**.

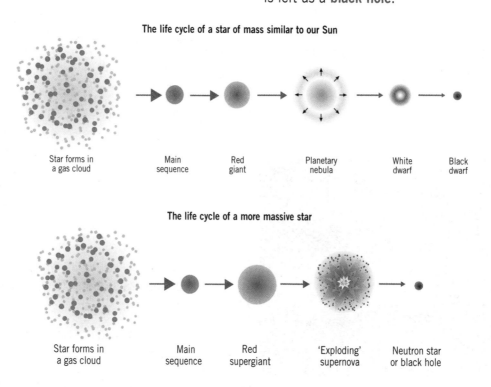

The life cycle of a star of mass similar to our Sun

| Star forms in a gas cloud | Main sequence | Red giant | Planetary nebula | White dwarf | Black dwarf |

The life cycle of a more massive star

| Star forms in a gas cloud | Main sequence | Red supergiant | 'Exploding' supernova | Neutron star or black hole |

Build Your Understanding

The particles of dust and gases in the interstellar gas clouds are attracted by gravitational forces between the particles. This is why the cloud contracts. As the core gets hotter the atoms collide at high speed, losing their electrons. The temperature has to be high enough to force the nuclei close together for fusion to occur. When light nuclei fuse they release energy.

In a main sequence star the high pressure in the core is balanced by the gravitational forces. The largest stars fuse hydrogen the most quickly so, surprisingly, more massive stars have shorter lifetimes.

Only the most massive stars can cause larger nuclei to fuse, forming elements like magnesium and silicon, but even they cannot form nuclei more massive than iron.

Black holes are so dense that even light cannot escape from their strong gravitational fields.

✓ Maximise Your Marks

Learn the names of all the different stages in the 'lifetime' of a star and make sure that you can use them correctly to describe what happens at each stage.

⚲ Boost Your Memory

To remember the stages in the lifetime of a star use a mnemonic, for example:

New play mates say red giants play nicely with dwarfs.

Nebula, protostar, main sequence, red giant, planetary nebula, white dwarf.

How we know about Stars

All the information we have about objects outside the Solar System comes from observations made with telescopes. What we know depends on the electromagnetic radiation from the stars and galaxies.

All around the Universe there are new stars being formed, main sequence stars in the stable part of their 'lifetime' and older stars coming to the end of their time as a star. By observing all of these scientists have worked out the 'life history' of stars.

To find out what stars are made of scientists look at the spectral lines (see page 50) in the spectrum of radiation from a star. These are used to identify the elements present in the star.

❓ Test Yourself

1 What is a planetary nebula?

2 When our Sun has completed the whole 'lifetime', what will be left at the end?

3 Which is brighter, a new main sequence star, a new white dwarf or a new supernova?

4 What is the difference between a red giant and a red supergiant star?

★ Stretch Yourself

1 How were elements like the carbon and oxygen in your body formed?

2 Why is a black hole black?

Galaxies and Red-Shift

Beyond the Earth

Galaxies

Our Sun is a star in the **Milky Way galaxy**. To see the Milky Way from Earth (it looks like a milky strip of stars in the sky) you need to be far away from the light pollution of towns and cities.

Galaxies are collections of thousands of millions of stars. There are thousands of millions of galaxies in the Universe. All the galaxies are moving away from us, and from each other.

Build Your Understanding

The Milky Way galaxy is shaped like a flat disc with spiral arms. Between galaxies there are empty regions that are hundreds of millions of light years across.

⚡ Boost Your Memory

Remember 'thousands of millions' of stars in a galaxy *and* galaxies in the Universe.

How Science Works

The **scientific community** shares ideas in meetings called **conferences** and in **published journals**. They evaluate each other's work. This process is called **peer review**.

In 1920 scientists had questions about our galaxy and the Universe:
- What were **spiral nebulae**?
- Was our galaxy, the Milky Way, the only galaxy?
- How big was the Milky Way galaxy?

Two scientists held a **Great Debate**.

Harlow Shapley argued that the Milky Way was the only galaxy. He thought spiral nebulae were gas clouds. Heber Curtis argued that spiral nebulae were other galaxies at great distances away from the Milky Way.

Evidence

There was not enough evidence to prove who was right. At the time most scientists thought that Shapley made a better case.

In 1924 Edwin Hubble used a new telescope and a new method of measuring distances. This new evidence showed that the distances to spiral nebulae were much greater than the size of the Milky Way.

He published his results and they were **evaluated** and **reproduced** by other scientists. Distances to more spiral nebulae were measured. They were outside the Milky Way. The scientific community accepted Curtis was right.

New Explanations

When new data disagrees with an accepted explanation, scientists do not immediately give up their accepted explanation. They wait until they have an explanation that fits the data better before they give up the accepted explanation.

Absorption Spectra

Spectral lines are lines in the spectrum of radiation from a hot object like a star.

Dark lines are wavelengths that have been **absorbed** by the atoms of gases the radiation has passed through, usually in the outer, cooler parts of the stars. Bright lines are wavelengths **emitted** by atoms of hot gases.

Each **element** has a different set of spectral lines, so the lines can be used to identify the elements present in the stars. The **absorption spectrum** has been described as a 'fingerprint' of the element.

Red-shift

For visible light, red has the longest wavelength and violet the shortest. If the wavelength is longer than expected this is called a **red-shift**.

If a source of waves is moving away the wavelength appears longer.

A red-shift in the light from a star shows that the distance between us and the star is increasing. The bigger the red-shift the faster the star is moving away.

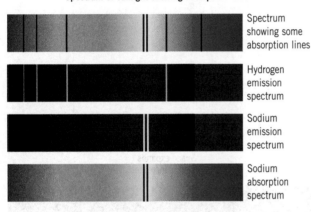

Spectrum of sunlight showing absorption lines

Spectrum showing some absorption lines

Hydrogen emission spectrum

Sodium emission spectrum

Sodium absorption spectrum

Hubble discovered:
- The light from all the distant galaxies is red-shifted – so they are all moving away from us.
- The further away the galaxy, the bigger the red-shift, so the faster the galaxy is moving away.

Waves from a moving source

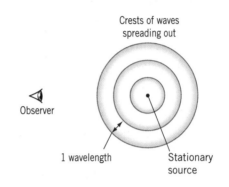

Crests of waves spreading out

Observer

1 wavelength

Stationary source

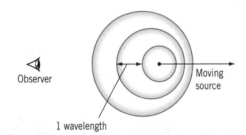

Observer

1 wavelength

Moving source

Build Your Understanding

The change in wavelength, because a source of waves is travelling towards or away from an observer, is called the Doppler Effect. When the wavelength increases the frequency decreases. This is why a siren drops in frequency as it travels away and increases in frequency as it approaches.

Light from stars and galaxies is red-shifted or blue-shifted depending on whether they are moving towards us or away from us. The shift moves all the dark spectral lines by the same amount.

✓ Maximise Your Marks

A red-shift is a shift towards the red end of the spectrum. Infrared radiation would be shifted towards microwaves not towards red light. A blue-shift means the radiation source is moving towards the observer.

? Test Yourself

1. What do we know about the movement of distant galaxies?

2. What is peer review?

3. From the absorption spectra, is sodium vapour present in the Sun?

4. If the Sun was moving away from us which way would the spectral lines move?

★ Stretch Yourself

1. Light from the Andromeda galaxy is blue-shifted. What does this tell you about the galaxy?

2. What information can astronomers get from line spectra?

Expanding Universe and Big Bang

The Expanding Universe

Hubble made two important observations:
- The light from all the distant galaxies is red-shifted.
- The further away the galaxy the bigger the red-shift.

This means:
- All the distant galaxies are moving away from us.
- The further away the galaxy, the faster it is moving away.

We would not see these patterns in the red-shifts just because we, or the galaxies, are moving through space, but it is what we would see if space was expanding.

This is why scientists think we live in an **expanding Universe**. The Universe is everything that exists. There is nothing outside the Universe – not even empty space.

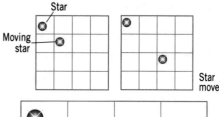
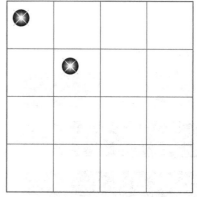

The difference between a moving star and expanding space

Star

Moving star

Star moves

Space expands distance increases

ⓘ Boost Your Memory

To help you to remember about space expanding and the two effects that Hubble observed, imagine blowing-up a balloon with galaxies drawn on it. The galaxies drawn on the balloon will all spread out, moving away from each other and those furthest apart will separate more quickly because the plastic surface is expanding.

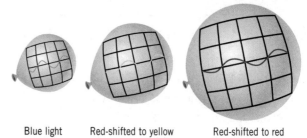

Blue light   Red-shifted to yellow   Red-shifted to red

Build Your Understanding

All of the space in the Universe is expanding, but we do not notice this in the Solar System. This is because gravity is an attractive force that stops mass spreading out when space expands.

Stars and the nearest galaxies may show red-shifts or blue-shifts because they are moving through space and these are sometimes called Doppler shifts. Distant galaxies show red-shifts because space is expanding, so these are sometimes called cosmological red-shifts.

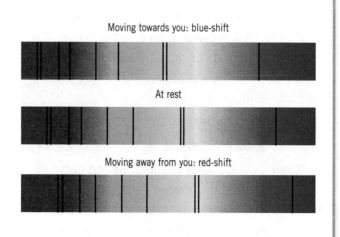

Moving towards you: blue-shift

At rest

Moving away from you: red-shift

The Big Bang

The fact that the Universe is expanding is strong evidence for the fact that it started as a small point. Starting from today and working backwards scientists calculate that the Universe began with a 'Big Bang' about 14 thousand million years ago. This is called the **Big Bang Theory**.

The Big Bang Theory started as a hypothesis – a suggested explanation thought up creatively to account for the data. Scientists then used it to make a prediction. They said that the Big Bang would have produced radiation that, by now, would be found in the microwave region of the spectrum. It would come from all parts of the Universe. Scientists called this the **cosmic microwave background radiation (CMBR)**.

Build Your Understanding

Scientists began searching for the CMBR and in 1965 two scientists, Arno Penzias and Robert Wilson, discovered it accidently.

They were using a radio telescope and could not account for an annoying microwave signal that seemed to come equally from all directions. It was the CMBR.

The Big Bang Theory was the only theory that could account for it, so this evidence led to the theory being accepted by most scientists.

NASA image of the CMBR

Questions that can't be answered

Some questions like 'What happened before the Big Bang?' scientists may never be able to answer because they can't collect data from before the Big Bang.

Some questions scientists can't answer because of difficulties in collecting the data they need. For example, whether the Universe will continue to expand, or whether it will slow down and stop, or whether it will start to contract, depends on the mass of the Universe and the distance to the edge of the Universe. These are very large and difficult to measure, so scientists cannot yet answer the question, 'What will be the ultimate fate of the Universe?'

The Steady State Theory

The **Steady State Theory** is an alternative theory that suggests the Universe is expanding, but has always looked the same. It will continue to look the same, because new matter is being created at various places. There were already some problems with data that could not be explained by the Steady State Theory and the discovery of the CMBR led to the general acceptance of the Big Bang Theory.

✓ Maximise Your Marks

Do not confuse the CMBR with radioactive background radiation, which is ionising radiation from radioactive materials.

❓ Test Yourself

1. What explanation do scientists have for the red-shifts in radiation from all the distant galaxies?
2. How long ago was the Big Bang?
3. What did the Big Bang Theory predict that scientists started to search for?
4. Give an example of a question science can't answer.

⭐ Stretch Yourself

1. What effect was predicted by scientists if the Big Bang theory was true?
2. What led to the acceptance of the Big Bang Theory and why?

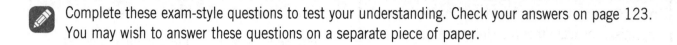

Practice Questions

Complete these exam-style questions to test your understanding. Check your answers on page 123. You may wish to answer these questions on a separate piece of paper.

1 Where is the asteroid belt found? Why have the asteroids not clumped together to form a planet?

.... (2)

2 Label this diagram of the Solar System.
Draw an X where you would find the asteroid belt. (2)

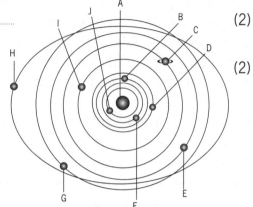

3 The following statements about space travel are reported in the press. Write **T** for the statements that are **true** and **F** for the statements that are **false**. (3)

a) A manned space trip to the nearest star would take four years. ....

b) People living on the Moon would experience no gravity, so they would lose bone mass and their muscles would waste away. ....

c) A spacecraft would need to be shielded from cosmic rays as the radiation is dangerous for humans. ....

d) Unmanned spacecraft can use remote sensors to send back information about temperature, the magnetic field, gases in the atmosphere and photographs. ....

e) Unmanned missions to Mars have sent back a lot of information. ....

f) A spaceship will require heating to keep astronauts warm. ....

4 Put these objects in order of size, starting with the smallest.

Jupiter   Milky Way   Pluto   Solar System   Earth's orbit   Sun

.... (3)

5 From their observations, astronomers think that two stars Alpha and Beta are the same type of star. They say that star Alpha is twice as far away as star Beta.

a) What would you expect to see when you compare star Alpha with star Beta?

.... (1)

b) In six months one of the stars appears to move further against the background of distant stars than the other. Which star appears to move more?

.... (1)

6 Inside stars energy is released by: (1)

    A Helium nuclei fusing to give hydrogen nuclei.

    B Helium nuclei splitting (fission) to give hydrogen nuclei.

    C Hydrogen nuclei fusing to give helium nuclei.

    D Hydrogen nuclei splitting (fission) to give helium nuclei.

7 Which of these best describes the way distant galaxies move? (1)

    A Distant galaxies are stationary, but our galaxy is moving away from them.

    B Distant galaxies are all moving away from each other at constant speed.

    C Distant galaxies are all moving away from each other. Those furthest away are moving fastest.

    D Some distant galaxies are moving away from our galaxy and some are moving towards it.

8 When we observe light from a distant star we see a red-shift in the wavelength. What does it tell us about the star? (1)

9 The table lists the wavelengths of the absorption spectral lines in the spectrum of a star and the spectral lines of some elements.

Wavelength of absorption spectral lines of star (nm)	Wavelength of spectral lines for elements (nm)			
Star	Helium	Hydrogen	Iron	Magnesium
486	502	486	467	517
502	588	656	527	518
588	668			
589				
590				
656				
668				

Which elements are present in the star? (1)

10 Most scientist accept the theory called The Big Bang Theory to explain the beginning of the Universe. Explain what this theory is and the evidence that scientists have for accepting it. (In your answer, aim to write between 8 and 12 lines.) (6)

How well did you do?

| 0–7 | Try again | 8–12 | Getting there | 13–17 | Good work | 18–22 | Excellent! |

Distance, Speed and Velocity

Distance, Speed and Velocity

Speed is measured in metres per second (m/s) or kilometres per hour (km/h):

- An athlete running with a speed of 5 m/s travels a distance of 5 metres in one second and 10 metres in two seconds.
- An athlete with a faster speed of 8 m/s travels further, 8 metres, in each second and takes less time to complete his journey.

To calculate speed:

$$\text{speed (m/s)} = \frac{\text{distance (m)}}{\text{time (s)}}$$

⚡ Boost Your Memory

Some students find using the following triangle method useful to rearrange a formula.

Example: speed $= \dfrac{\text{distance}}{\text{time}}$

To calculate distance or time:

- Write the formula into a triangle, so that distance is 'over' time. This means putting distance at the top, time can go in either of the other corners.

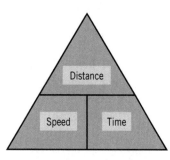

- To find the **distance**, cover the word distance with your finger and look at the position of speed and time. They are side by side, so **distance = speed × time**.
- To find the **time**, cover the word time with your finger and distance is 'over' speed, so

$$\text{time} = \frac{\text{distance}}{\text{speed}}$$

Build Your Understanding

There are two ways of looking at a journey:

- The distance travelled can only increase, or stay the same, so speed is always a positive number.
- The direction travelled is important, so that travel in one direction is a positive distance and in the opposite direction is a negative distance. Sometimes, distance in a given direction is called displacement.

Quantities that have magnitude and direction are called vectors.

Velocity is a vector, because velocity is speed in a given direction. For example, a dog walks in a positive direction and then back again with a constant speed of 2 m/s, so he walks with a velocity of +2 m/s and then with a velocity of –2 m/s.

$$\text{velocity (m/s)} = \frac{\text{displacement (m)}}{\text{time (s)}}$$

✓ Maximise Your Marks

When you are doing calculations:

- Always show your working – if you make a mistake in the calculation you may still get some marks.
- Do not write down *only* the triangle. You will not get marks for using the correct formula or equation.

Distance-Time Graphs

On a **distance-time graph**:

- A horizontal line means the object is stopped.
- A straight line sloping upwards means it has a steady speed.

The steepness, or **gradient**, of the line shows the speed:

- A steeper gradient means a higher speed.
- A curved line means the speed is changing.

Distance-Time Graphs (cont.)

Example: Distance-time graph for a cycle ride.

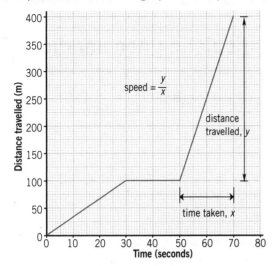

Between 30 s and 50 s the cyclist stopped. The graph has a steeper gradient between 50 s and 70 s than between 0 s and 30 s. The cyclist was travelling at a greater speed.

To calculate a speed from a graph, work out the gradient of the straight line section.

speed = y/x where y = 400 m −100 m = 300 m and x = 70 s − 50 s = 20 s.

$$\text{speed} = \frac{300 \text{ m}}{20 \text{ s}} = 15 \text{ m/s}.$$

Build Your Understanding

If the direction of travel is being considered:
- A negative displacement is in the opposite direction to a positive displacement.
- A straight line sloping upwards or downwards means steady speed.
- Upwards means a steady positive velocity, and downwards means a steady negative velocity.

Example: Displacement-time graph for a journey from home.

A boy starts from home (0 km) and walks to a shop, home again and then in the opposite direction to the shop.

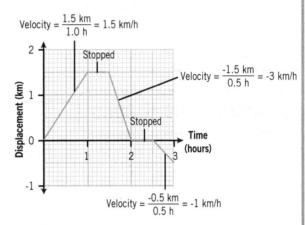

Average Speed and Instantaneous Speed

The **average speed** of the cyclist for the total journey shown on the graph is:

$$= \frac{\text{total distance}}{\text{total time}} = \frac{400 \text{ m}}{70 \text{ s}} = 7.62 \text{ m/s}$$

This is not the same as the **instantaneous speed** at any moment because the speed changes during the journey. If you calculate the average speed over a shorter time interval you get closer to the instantaneous speed.

? Test Yourself

1. What is the difference between instantaneous speed and average speed?

2. In the graph of the cycle ride, what is the speed during the first 30 seconds?

3. A car travels 288 km in three hours. Calculate the speed in km/h.

4. A car travels at a speed of 12 m/s. How long will it take to travel 1.44 km?

★ Stretch Yourself

1. In the graph of the journey from home:

   a) What was the displacement after the first 36 minutes?

   b) When was the speed greatest?

   c) How can you tell this without doing any calculations?

Speed, Velocity and Acceleration

Acceleration

A change of velocity is called **acceleration**. Speeding up, slowing down and changing direction are all examples of acceleration.

Acceleration is the change in velocity per second. It is measured in metres per second squared (m/s^2).

Example: If a car accelerates from 0 to 27 m/s (about 60 mph) in six seconds the change in velocity is 27 m/s, the acceleration:

$$= \frac{27 \text{ m/s}}{6 \text{ s}} = 4.5 \text{ m/s}^2$$

Calculating Acceleration

$$\text{acceleration (m/s}^2) = \frac{\text{change in velocity}}{\text{time taken (s)}} \text{ (m/s)}$$

$$a = \frac{(v-u)}{t}$$

Where a is the acceleration of an object whose velocity changes from initial velocity u to final velocity, v in time t.

Example: A car accelerates from 14 m/s to 30 m/s in 8 s. The acceleration:

$$a = \frac{(30 \text{ m/s} - 14 \text{ m/s})}{8 \text{ s}} = \frac{16 \text{ m/s}}{8 \text{ s}} = 2 \text{ m/s}^2$$

Speed-Time Graphs

Plotting the speed of an object against the time gives a graph like this:

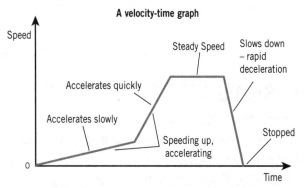

A velocity-time graph

- A **positive slope (gradient)** means that the speed is increasing – the object is accelerating.
- A **horizontal line** means that the object is travelling at a steady speed.

- A **negative slope (gradient)** means the speed is decreasing – negative acceleration.
- A **curved slope** means that the acceleration is changing – the object has **non-uniform acceleration**.

Tachographs are instruments that are put in lorry cabs to check that the lorry has not exceeded the speed limit and that the driver has stopped for breaks. They draw a graph of the speed against time for the lorry.

✓ Maximise Your Marks

Always check carefully whether a graph is a speed-time graph or a distance-time graph.

Build Your Understanding

On true speed-time graphs the speed has only positive values. On velocity-time graphs the velocity can be negative.

Graphs for a ball that rolls up a hill, slows and rolls back down, speeding up

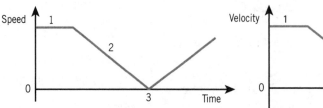
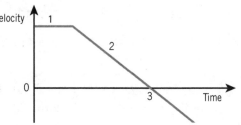

Forces and Motion

Acceleration from a Graph

The acceleration is the gradient of a velocity-time graph.

Example: A graph of velocity against time for a car journey.

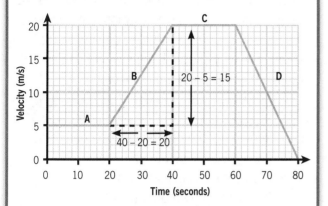

In part B of the graph the car accelerates from 5 m/s to 20 m/s in (40 − 20)s = 20 seconds. The gradient:

$$= \frac{y}{x} = \frac{15}{20} = 0.75$$

So the acceleration = 0.75 m/s².

⚡ Boost Your Memory

Remember the gradient is: $\dfrac{\text{rise}}{\text{run}}$

Rise

Run

Build Your Understanding

On a velocity-time graph the area between the graph and the time axis represents the distance travelled.

Example: Distance travelled on a bike journey.

The distance travelled: **= area under A + area under B**

Area A = area of pink triangle = ½ (10 m/s × 20 s) = 100 m

Area B = area of blue rectangle = 10 m/s × (60 − 20)s = 400 m

Distance travelled = 100 m + 400 m = 500 m

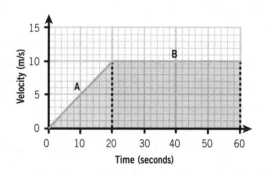

✔ Maximise Your Marks

Take care with units. You may need to change minutes or hours to seconds, or kilometres to metres:

$$36 \text{ km/h} = \frac{36 \times 1000}{60 \times 60} \text{ m/s} = 10 \text{ m/s}$$

Also remember to write down the units for your answer.

❓ Test Yourself

1. A car accelerates from 0 to 24 m/s in 8 s. What is the acceleration?

2. What does a horizontal line on a speed-time graph tell you?

3. From the graph of the car journey, what is the acceleration:

   a) in part C

   b) in part D?

⭐ Stretch Yourself

1. Sketch a velocity time graph for a car that speeds up in a negative direction, travels at a steady speed, and then slows down and stops.

2. From the graph of the car journey, what is the distance travelled in:

   a) in part C

   b) in part D?

Forces

Mass and Weight

Mass is measured in kilograms (kg). It is the amount of matter in an object. An object has the same mass everywhere, on the Earth, the Moon, or in outer space.

Weight is a force and is measured in newtons (N). Weight is the force of gravity attracting the mass towards the centre of the Earth:

weight (N) = mass (kg) × gravitational field strength (N/kg)

Build Your Understanding

In outer space there is no gravity so all objects are weightless. The Moon's gravitational attraction is only one sixth of the Earth's, so objects on the Moon have only one sixth their weight on Earth, though their mass remains constant.

Boost Your Memory

To remember the differences between mass and weight think of a tin of beans:
- It's weightless in outer space.
- It has less weight on the Moon.
- Its mass changes if you eat the beans.

Resultant and Balanced Forces

Forces have size and direction. The length of the arrow on a diagram represents the size of the force.

When several forces act on an object the effect is the same as one force in a certain direction. This is called the resultant force. Forces combine to give a resultant force. If the resultant force is zero the forces on the object are balanced.

A resultant force changes the velocity of an object. This idea is known as Newton's First Law of Motion:

If the resultant force on an object is zero, the object will remain stationary or continue to move at a steady speed in the same direction.

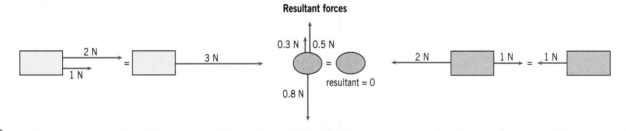

Resultant forces

More Balanced Forces

When forces on an object are balanced it does not fall.

Reaction and tension are forces that can balance weight

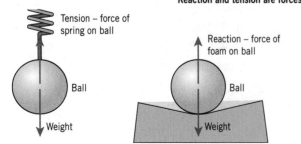

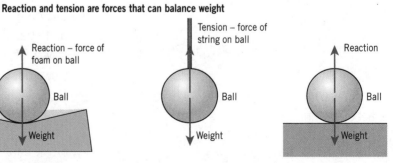

The upward force of the spring, or the foam, trying to return to their original shape balances the weight.

Even though the changes in shape are too small for us to see, the restoring force – the tension in the string or the reaction from the floor, balances the weight.

Resistance to Motion-Friction Forces

When one object slides, or tries to slide, over another, there is **friction**, the resistive force between the two surfaces.

Air resistance is the resistive force that acts against objects moving through the air.

Drag is the resistive force on objects moving through liquids or gases. Drag is larger in liquids.

These resistive forces:
- Always act against the direction of motion.
- Are zero when there is no movement.
- Increase as the speed of the object increases.

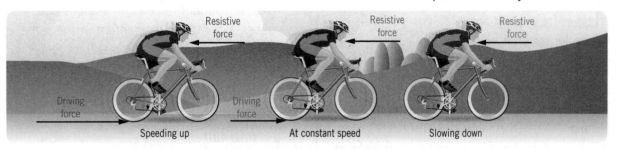

Speeding up | At constant speed | Slowing down

Forces when Falling

The resultant force on a skydiver changes.

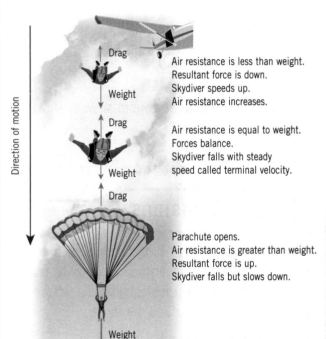

Air resistance is less than weight.
Resultant force is down.
Skydiver speeds up.
Air resistance increases.

Air resistance is equal to weight.
Forces balance.
Skydiver falls with steady speed called terminal velocity.

Parachute opens.
Air resistance is greater than weight.
Resultant force is up.
Skydiver falls but slows down.

Falling objects reach a steady speed called the terminal velocity, when: **drag = weight**.

In a vacuum, there is no air resistance so objects continue to fall with acceleration due to gravity of 10 m/s^2.

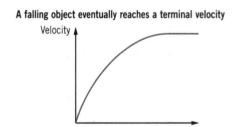

A falling object eventually reaches a terminal velocity

✓ Maximise Your Marks

Remember if an object is travelling at a steady speed there is no resultant force on it. When an object is slowing down the resultant force on it is opposite to its direction.

It's a common mistake to think that there is always a force in the direction of movement.

? Test Yourself

1. What is the weight of 2 kg of sugar?

2. The driving force is 3000 N and friction is 900 N. What is the resultant force?

3. A car travels with steady speed of 100 km/h in a straight line. What is the resultant force?

★ Stretch Yourself

1. How much does a 100 g apple weigh a) on Earth and b) on the Moon?

2. The driving force is 2500 N, friction is 800 N and air resistance is 1700 N. What is the resultant force?

Acceleration and Momentum

Force and Acceleration

A resultant force on an object causes it to accelerate. The acceleration is:
- Larger for a larger force.
- Smaller for a larger mass.
- In the same direction as the force.

For a resultant force on an object:
force (N) = mass (kg) × acceleration (m/s²)
$F = ma$
Where F = force, m = mass and a = acceleration.

Force and Momentum

The **momentum** of an object is:
- Larger for a larger velocity.
- Larger for a larger mass.
- In the same direction as the velocity.
- Measured in units called kg m/s or Ns (they are the same).

momentum (kgm/s or Ns) = mass (kg) × velocity (m/s)
where mv = momentum, v = velocity.

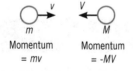

When a resultant force acts on an object, it causes a change in momentum in the same direction as the force (because the velocity changes).

The bigger the force and the longer the time that the force acts, the bigger the change in momentum:

change in momentum (kg m/s or Ns) = force (N) × time (s)

💡 Boost Your Memory

To remember what momentum depends on, think about 10 pin bowling. You are more likely to knock pins down with more momentum. This could be a ball with a lot of mass or a ball with a high velocity.

Build Your Understanding

Newton's Second Law of Motion states:

When a resultant force acts on an object, it causes a change in momentum in the same direction as the force. The resultant force equals the rate of change of momentum.

$$\text{force (N)} = \frac{\text{change in momentum}}{\text{time (s)}} \text{ (kgm/s or Ns)}$$

$$F = \frac{(mv-mu)}{t}$$

Where t = time, u = initial velocity and v = final velocity.

The equation $F = ma$ is another way of saying this because acceleration:

$$a = \frac{(v-u)}{t}$$

$$F = \frac{m(v-u)}{t} = \frac{(mv-mu)}{t}$$

You do not need to be able to show or explain this.

Safer Collisions

In a collision, a force brings your body to a sudden stop. The larger the stopping force on the body the more it is damaged. To reduce damage we must reduce the force:

change of momentum = force × time

For the same change in momentum, to reduce the force we must increase the time to stop.

If the collision takes place over a longer time, say 0.5 s instead of 0.05 s – 10 times as long – then the stopping force will only be one tenth of the size.

Stopping Safely

These safety features work by increasing the time taken for collision:

- **Crumple zones** in the car. The front and back of the car are designed to crumple in a collision, increasing the distance and time over which the occupants are brought to a stop.

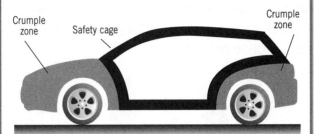

Crumple zone   Safety cage   Crumple zone

- The body hits the **airbag**, which is compressed, increasing the distance the body moves and the time it takes to stop.
- **Seatbelts** are designed to stretch slightly so that the body moves forward and comes to a stop more slowly with a smaller deceleration than it would if it hit the windscreen or front seats. After a collision the seatbelts should be replaced because they are not elastic and once stretched will not return to their shape and may break in a second collision.
- **Cycle and motorcycle helmets** contain a layer of material which will compress on impact so that the skull is brought to a stop more slowly. They should be replaced after a collision as the material will be damaged and may not protect you again.

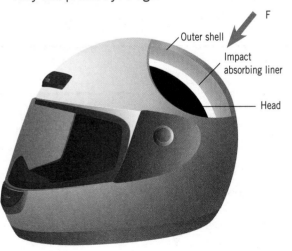

F

Outer shell

Impact absorbing liner

Head

Build Your Understanding

Other examples of reducing the force by increasing the time taken to stop include:
- Crash barriers that crumple on impact.
- Bending your knees when you land after jumping.
- Bubble wrap.
- Sprung floors in gyms.

✓ Maximise Your Marks

Do not confuse momentum with energy. Momentum has a direction and is measured in Ns or kg m/s. Energy, like mass, has no direction and it is measured in Joules (J). 1 J = 1 Nm – a different unit.

❓ Test Yourself

1. Calculate the resultant force on a 1200 kg car accelerating at 3 m/s^2.
2. Calculate the momentum of a ball of mass 2 kg and velocity 5 m/s.
3. Calculate the momentum of a ball of mass 200 g and velocity 8 m/s travelling in the opposite direction.
4. A force of 50 N acts on a stationary object for 12 seconds. Calculate its gain in momentum.

⭐ Stretch Yourself

1. A runaway truck mass 1000 kg and velocity 12 m/s came to a sudden stop in 0.002 s.

   a) Calculate the stopping force on the truck.

   b) A crash barrier would have stopped it over 0.5 s. What would the stopping force have been?

   c) What difference would this make?

Pairs of Forces: Action and Reaction

Interaction Pairs

When two objects interact there is always an **interaction pair of forces**. As these skaters show, the boy cannot push the girl without the girl pushing the boy.

In an interaction pair of forces, the two forces:
- Are always equal in size and opposite in direction.
- Always act on different objects.
- Are always the same type of force (for example, contact forces, gravitational forces, or magnetic forces).

This idea is known as **Newton's Third Law of Motion** which states:

When two objects interact, the forces they exert on each other are equal and opposite and are called action and reaction forces.

Push

Friction and Getting Started

Friction is the reaction force needed for walking or wheeled transport.

- **Action force** is where the wheel pushes back on the road.
- **Reaction force** is where the road pushes forward on the wheel, which sends the wheel forward.

> ### 💡 Boost Your Memory
>
> To remember how friction gets you moving, think of trying to cycle on a frictionless icy surface. There is no reaction force. You would slip backwards and never move forward.

When you walk your foot pushes back on the ground and the ground pushes your foot forward.

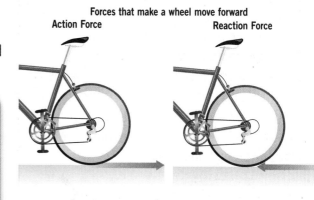

Forces that make a wheel move forward
Action Force   **Reaction Force**

Build Your Understanding

The weight of an object is the gravitational attraction towards the centre of the Earth. The other force of this interaction pair acts on the Earth. It is the gravitational attraction of the object attracting the whole Earth.

We do not notice this effect because the mass of the Earth is so large.

An action and reaction pair of forces

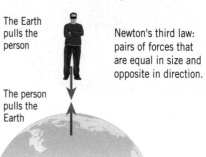

The Earth pulls the person

The person pulls the Earth

Newton's third law: pairs of forces that are equal in size and opposite in direction.

Different Forces

Do not confuse interaction pairs of forces, which act on different objects, with balanced forces, which act on the same object.

The reaction force from a surface *balances* the weight of the object. It is not the reaction pair of the weight, because:
- It is a different type of force (a contact force, not gravitational).
- It acts on the same object.

Rockets and Jet Engines

Rockets and jet engines both produce hot exhaust gases. These are pushed out of the back of the engines. There is an equal and opposite reaction force that sends the rocket or jet forward.

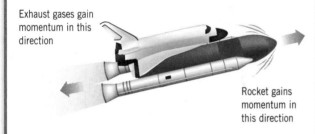

Exhaust gases gain momentum in this direction

Rocket gains momentum in this direction

Recoil

When a bullet leaves a gun, action and reaction, or **conservation of momentum**, tell us the gun must recoil.

Example: A 0.8 g paintball is fired at 80 m/s from a 3 kg paintball marker.

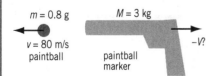

$m = 0.8$ g

$v = 80$ m/s
paintball

$M = 3$ kg

$-V$?
paintball marker

Conservation of momentum: $m v = M V$

$V = \dfrac{0.8 \text{ g} \times 80 \text{ m/s}}{3000 \text{ g}} = 0.021$ m/s

Build Your Understanding

When two objects, collide or explode apart there is an equal and opposite force on each object for the same time, so the change in the momentum of the objects is equal and opposite. Momentum is conserved. The total momentum of two objects before collision or explosion is the same as the total momentum after.

Example: When two objects collide and stick together.

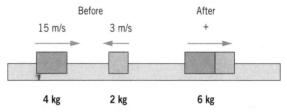

Before

15 m/s   3 m/s

After
+

4 kg   2 kg   6 kg

Before the collision momentum:

= 4 kg × 15 m/s + 2 kg × (–3) m/s

= (60 – 6) kg m/s = 54 kg m/s

After the collision: momentum = 6 kg × v

So, because of conservation of momentum:

$v = \dfrac{54}{6} = 9$ m/s

$v = 9$ m/s

❓ Test Yourself

1. A girl pushes on a wall with a force of 5 N. Describe the reaction force.

2. Why is it difficult to walk on ice?

3. How does a rocket move in outer space where there is nothing to push against to get moving?

4. When you release a partly inflated balloon it flies around as it deflates. Explain why.

⭐ Stretch Yourself

1. A book is placed on a table. What are the two interaction pairs of forces?

2. A toy car with mass 0.5 kg and speed 4 m/s collides with a toy truck of mass 2 kg. They both stop. What was the speed of the truck?

Work and Energy

Work and Energy

When a **force** makes something move **work** is done. The work done is equal to the energy transferred. Work and energy are measured in joules (J):

work done by a force (J) = force (N) × distance moved by force in direction of the force (m)

When work is done *by* something it loses energy, when work is done *on* something it gains energy.

Kinetic Energy

An object that is moving has **kinetic energy** (**KE**). The energy depends on the mass of the object and on the square of the speed. Doubling the speed gives four times the energy.

Example: An air hockey puck, floating on an air table, is almost frictionless. A force does work on the puck – it pushes it a small distance. Energy is transferred and the kinetic energy of the puck increases – it speeds up. When the force stops the puck moves at a constant speed across the table – its kinetic energy is now constant.

kinetic energy (J) = ½ × mass (kg) × [speed (m/s)]²

Energy does not have a direction. Speed or velocity can be used to calculate the kinetic energy.

Example: A ball of mass 0.27 kg and speed 3 m/s has KE = 0.5 × 0.27 kg × (3 m/s)² = 0.9 J.

Gravitational Potential Energy

Gravitational potential energy (**GPE**) is the stored energy that an object has because of its position above the surface of the Earth.

Example: Doing work – increasing the GPE.

When you lift a 10 N weight (a mass of 1 kg) from the floor to a high shelf, a height difference of 2 m, you have done work on the weight.

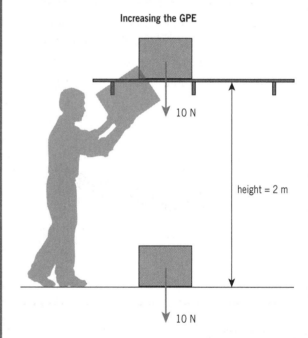

Increasing the GPE

10 N

height = 2 m

10 N

The work done = 10 N × 2 m = 20 J and this is equal to the increase in the GPE of the weight. **Change in GPE (J) = weight (N) × vertical height difference (m).**

💡 Boost Your Memory

It's the *change* in GPE that depends on the *change* in height.

Build Your Understanding

Change in **GPE = m g h** where g = the gravitational field strength (N/kg), m = mass (kg) and h = height change (m).

KE = ½ $m v^2$ where m = mass (kg) v = speed (m/s)

Rollercoasters

When frictional forces are small enough to be ignored the transfer of energy between KE and GPE can be used to calculate heights and speeds.

Example: A car is driven by a trackside motor to the top of a rollercoaster and then freewheels down the slope.

Increase in GPE of car:

$$= m\,g\,h = 1000\ \text{kg} \times 10\ \text{N/kg} \times 45\ \text{m} = 450\,000\ \text{J}$$

Assuming there are no friction forces as the train travels down the slope:

Loss of GPE = gain in KE = 450 000 J

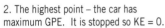

Transferring energy from GPE to KE and back again

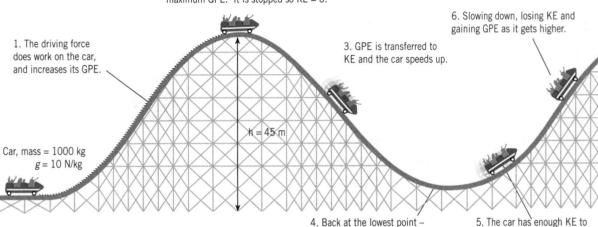

2. The highest point – the car has maximum GPE. It is stopped so KE = 0.

6. Slowing down, losing KE and gaining GPE as it gets higher.

1. The driving force does work on the car, and increases its GPE.

3. GPE is transferred to KE and the car speeds up.

$h = 45$ m

Car, mass = 1000 kg
$g = 10$ N/kg

4. Back at the lowest point – maximum KE and GPE = 0.

5. The car has enough KE to continue up the next slope.

Build Your Understanding

Example: To calculate the speed of a car as above:

Gain in KE = 450 000 J

$450\,000\ \text{J} = \tfrac{1}{2}\,m\,v^2 = \tfrac{1}{2} \times 1000\ \text{kg} \times v^2$

$v^2 = 900\ (\text{m/s})^2$ so speed $v = 30$ m/s

✓ Maximise Your Marks

Remember to square a number by multiplying it by itself. A calculator is useful for finding the square root of a number.

❓ Test Yourself

1. What is the work done by a tractor that pulls a trailer with a force of 1000 N across a 200 m field?

2. A rollercoaster car that has a weight of 12000 N goes to the top of a 30 m slope. What is the gain in GPE?

3. A 900 kg car is travelling at 15 m/s. What is its KE?

4. A toy car rolls down a slope. The gain in KE is less than the loss in PE. Suggest why.

⭐ Stretch Yourself

1. A lift weighs 8000 N it is raised a height of 50 m.

   a) What is the gain in GPE?

   b) What is the work done on the lift by the motor?

Energy and Power

Conservation of Energy

The **Principle of Conservation of Energy** says that the total energy always remains the same. When energy is transferred to the surroundings by heating due to frictional forces it is no longer useful, but it is not lost. We say it has been **dissipated** (spread out) as heat.

The relationship 'gain in KE = loss in GPE' is only true for a falling object if the air resistance (or drag) is small and can be ignored, or if the object is falling in a vacuum.

A skydiver eventually reaches **terminal velocity**. She is still falling, so GPE is being lost, but no KE is being gained. The energy is being used to do work against the frictional force (air resistance). The skydiver and surrounding air will heat up.

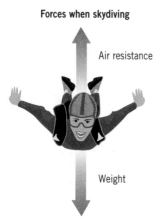

Forces when skydiving

Air resistance

Weight

Build Your Understanding

The space shuttle, with a lot of KE, needed heat proof tiles to protect it from the heat resulting from doing work against air resistance when it re-entered the Earth's atmosphere.

When a cyclist pedals, but travels at a steady speed, work is done against air resistance and friction. Energy is transferred and heats the bicycle and surroundings. No energy is being transferred as KE to the bicycle unless it speeds up.

Stopping Distances

The distance travelled between the driver noticing a hazard and the vehicle being stationary is called the **stopping distance**:
- Stopping distance = thinking distance + braking distance.
- **Thinking distance** is distance travelled during the drivers reaction time – the time between seeing the hazard and applying the brakes.

- **Braking distance** is the distance travelled while the vehicle is braking.

This diagram shows the shortest stopping distances at different speeds.

When speed doubles:
- Thinking distance doubles.
- Braking distance is four times as far.

The stopping distance increases with speed

30 mph

9 m   14 m

60 mph

18 m   56 m

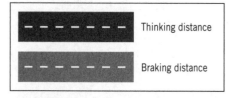

Thinking distance

Braking distance

Longer Stopping Distances

The stopping distances are also longer if:

- The driver is tired, affected by some drugs (including alcohol and some medicines), or distracted and not concentrating. Thinking distance is increased.
- The road is wet or icy or the tyres or brakes are in poor condition. The friction forces will be less so the braking distance is increased.
- The vehicle is fully loaded with passengers or goods. The extra mass reduces the deceleration during braking, so the braking distance is increased.

These stopping distances are taken into account when setting road speed limits. Drivers should not drive closer than the thinking distance to the car in front, to allow for time to react. They should reduce speed in bad weather to allow for the increased braking distance.

Braking and Kinetic Energy

- When speed doubles reaction time is the same:
  thinking distance = speed × reaction time
  (The thinking distance doubles).
- Work done by the brakes against friction = loss in KE.

  braking force × braking distance = ½ mv^2

  braking distance = $\dfrac{\text{mass} \times \text{speed}^2}{2 \times \text{braking force}}$

  (The braking distance is four times as far.)

- The **thinking distance** depends on **speed**.
- The **braking distance** depends on **(speed)^2**.

  Example: At three times the speed, braking distance is nine times as far.

Power

Power is the work done, or energy transferred, divided by time. Power is measured in watts (W):

power (W) = $\dfrac{\text{work done or energy transferred (J)}}{\text{time (s)}}$

Build Your Understanding

Power is the rate of energy transfer.

Example: A 7.5 kW crane lifts a 3000 N weight up a height of 10 m. How long does it take?

$$7.5 \text{ kW} = \frac{3000 \text{ N} \times 10 \text{ m}}{t}$$

$$t = \frac{30000 \text{ J}}{7500 \text{ W}} = 4 \text{ s}$$

💡 Boost Your Memory

Units can help you to remember how things are related and do calculations. A watt is a joule per second, so divide energy (joules) by time (seconds) to get power (watts).

✓ Maximise Your Marks

Don't forget to use the correct units in your calculations. It can help to change everything to the basic units, for example kilowatts to watts. An answer without units is not an answer – it's just a number.

? Test Yourself

1. Under what conditions is the 'loss in GPE = gain in KE'?
2. What happens to the GPE when an object falls with terminal velocity?
3. Give an example where stopping distances would be longer than shown in the diagram.

★ Stretch Yourself

1. What happens to the energy transferred by a pedalling cyclist when travelling at a steady speed?
2. Calculate the stopping distance of a car when travelling at 90 mph.

Practice Questions

Complete these exam-style questions to test your understanding. Check your answers on pages 123–24. You may wish to answer these questions on a separate piece of paper.

1 This graph shows the movement of a car.

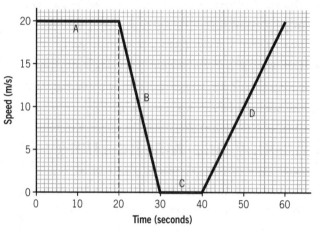

a) In which section, **A B C** or **D** is the car i) accelerating and ii) stopped? (2)

....

b) How far does the car travel in the first 20 seconds? (1)

....

c) What is the instantaneous speed of the car after 50 seconds? (1)

....

2 The graph below shows speed against time for a car journey.

a) Describe the motion of the car during the journey. (3)

....

b) What is:

i) The acceleration during the first 10 seconds? ....

ii) The speed after 20 s? ....

iii) The acceleration in the last 10 seconds? .... (2)

c) What does the shaded area represent? (1)

d) How far does the car travel:

i) In the first 10 seconds?

....

ii) In the first 30 seconds? (2)

3 Look at the table below.

Conditions (speed and seatbelt)	Time the body takes to stop (s)	Momentum of 80 kg body (kg m/s)	Stopping force on body (kN)
Seatbelt at 30 mph (13 m/s)	0.069	A	E
No seatbelt at 30 mph (13 m/s)	0.009	B	F
Seatbelt at 70 mph (31 m/s)	0.029	C	G
No seatbelt at 70 mph (31 m/s)	0.004	D	H

a) Calculate the missing values A-H to complete the table.
   There is a correlation between the speed a car is travelling and the severity of injuries in
   a collision. (3)

b) Is higher speed causing more serious injuries or is the correlation a coincidence? Explain your
   answer using data from the table. (3)

c) Injuries are reduced when people wear seatbelts. Use data from the table to explain why. (3)

d) Some people choose not to wear seatbelts. Suggest some risks and benefits of wearing seatbelts.
   (4)

4 Match the values of momentum to the situations. The car is travelling in the positive direction. (4)

Situation			Total momentum	
A	Car 1000 kg: +28 m/s	1	0 kg m/s	
B	Lorry 2800 kg: −10 m/s travelling towards car 1000 kg: +28 m/s	2	28000 kg m/s	
C	Car 1400 kg: −10 m/s travelling towards lorry 2800 kg: + 10 m/s	3	38000 kg m/s	
D	Lorry 2800 kg: +10 m/s travelling towards car 1000 kg: +10 m/s	4	14000 kg m/s	

5 When a fast car comes to a sudden stop in a collision the driver may be killed or severly injured.
   Describe some of the safety features of modern cars that reduce the risk and explain how they
   work. (In your answer, aim to write about 8 and 12 lines.) (6)

How well did you do?

| 0–7 | Try again | 8–16 | Getting there | 17–26 | Good work | 27–35 | Excellent! |

Electrostatic Effects

Electric Charge

Electric charge can be **positive** or **negative**. **Electrons** are particles with a negative electric charge. They can move freely through a **conductor**, for example any type of metal, but cannot move through an **insulator**.

Two objects attract each other if one is positively charged and the other is negatively charged. Two objects with similar charge (both positive or both negative) repel.

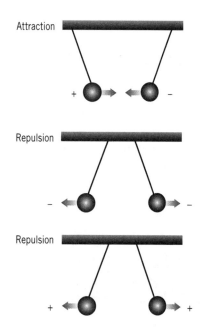

Electrostatic effects are caused by the transfer of electrons. (It is also sometimes called static electricity). When insulators are rubbed electrons are rubbed off one material and transferred to the other.

A polythene rod rubbed with a duster picks up electrons from the duster and becomes negatively charged, leaving the duster positively charged.

A Perspex rod rubbed with a duster loses electrons to the duster and becomes positively charged, leaving the duster negatively charged.

Materials that are positively charged have missing electrons. Materials that are negatively charged have extra electrons.

The insulated rod and the cloth have opposite charge

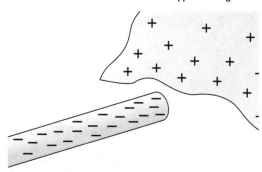

💡 Boost Your Memory

Like charges repel. Unlike charges attract. Remember, opposites attract!

Build Your Understanding

Conductors cannot be charged unless they are completely surrounded by insulating materials, such as dry air and plastic, otherwise the electrons flow to or from the conductor to discharge it.

An insulated conductor can be charged by rubbing it with a charged duster, or touching it with a charged rod. Some electrons are transferred, so that the charge is spread out over both objects.

A conductor can be discharged by touching it with another conductor, for example a wire, so that electrons can flow along the wire and cancel out the charge.

✓ Maximise Your Marks

Remember that it is the electrons that move from one object to another.

The Earth Connection

To stop conductors becoming charged they can be **earthed**. A thick metal wire is used to connect them to a large metal plate in the ground. This acts as a large reservoir of electrons. Electrons flow so quickly to, or from, earth that objects connected to earth do not become charged.

Dangerous or Annoying?

The human body conducts electricity. When a large flow of charge affects our nerves and muscles we call this an **electric shock**.

- Small electrostatic shocks are not harmful.
- Larger shocks can be dangerous to people with heart problems because a flow of charge through the body can stop the heart.
- Lightning is a very large electrostatic discharge. When it flows through a body it is often fatal.

Annoying Electrostatic Charge

Charged objects, like plastic cases and TV monitors, attract small particles of dust and dirt.

Clothing can be charged as you move and 'clings' to other items of clothing. Synthetic fibres are affected more than natural fibres as they are better insulators.

If charge builds up on you and you touch metal, for example a car door, the charge flows from you to the metal, and you get a shock.

A **spark** occurs when electrons jump across a gap. This can cause an explosion if there are:

inflammable vapours like petrol or methanol

powders in the air, like flour or custard, which contain lots of oxygen – as a dust they can explode

inflammable gases like hydrogen or methane

Build Your Understanding

Lorries containing flammable gases, liquids and powders are connected to an earth before loading or unloading. Aircraft are earthed before being refuelled. This prevents charge from building up on metal pipes or tanks when the loads are moved, so there is no danger of a spark igniting the load.

Anti-static sprays, liquids and cloths stop the build up of static charge. These work by increasing the amount of conduction, sometimes by attracting moisture because water conducts electricity.

If you stand on an insulating mat, or wear shoes with insulating soles, when you touch a charged object this will reduce the chance of an electric shock because the charge will not flow through you to earth. You become charged and stay charged until you touch a conductor.

? Test Yourself

1. Why is a plastic rod attracted to a cloth it has been rubbed with?

2. What particles are transferred when a balloon is rubbed with a cloth?

3. How many types of electric charge are there?

4. Why do you become charged walking on a nylon carpet, but not on a woollen carpet?

★ Stretch Yourself

1. When a plastic rod and a Perspex rod are both charged by rubbing they attract.

   a) What does this tell you about the charges on them?

   b) Would you expect two rubbed polythene rods to attract or repel?

2. Why are aircraft earthed before being refuelled?

Uses of Electrostatics

Electrostatic Precipitators

Electrostatic precipitators remove dust or smoke particles from chimneys, so that they are not carried out of the chimney by the hot air:

- Charged metal grids are put in the chimneys.
- The smoke particles pass through the grids and become charged.
- Plates at the side are charged opposite to the grids.
- The smoke particles are attracted and stick to the plates.
- The smoke particles clump together on the plates to form larger particles.
- The plates are struck and the large particles fall back down the chimney into containers.

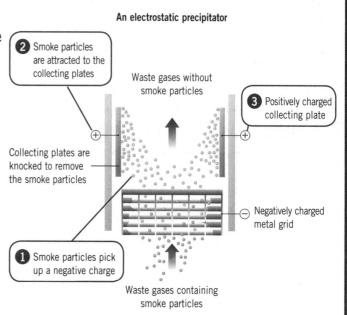

An electrostatic precipitator

② Smoke particles are attracted to the collecting plates

Waste gases without smoke particles

③ Positively charged collecting plate

Collecting plates are knocked to remove the smoke particles

Negatively charged metal grid

① Smoke particles pick up a negative charge

Waste gases containing smoke particles

Build Your Understanding

The grids are connected to a high voltage. They attract or repel charges in the smoke particles, so the particles become charged.

The grids are positively charged in some designs and negatively charged in others. If the grids are positively charged, the plates are earthed. The smoke or dust particles lose electrons and become positively charged. They induce a negative charge on the earthed metal plate and are attracted to the plate.

If the grids are negatively charged the plates are positively charged. The smoke or dust particles gain electrons and become negatively charged. They are attracted to the positively charged metal plates.

Paint Spraying

The paint and the object are given a different charge so that the paint is attracted to the object:

- The spray gun is charged so that it charges the paint particles.
- The paint particles repel each other to give a fine spray.
- The object is charged with the opposite charge to the paint.
- The object attracts the paint.
- The paint makes an even coat, it even gets underneath and into parts that are in shadow.
- Less paint is wasted.

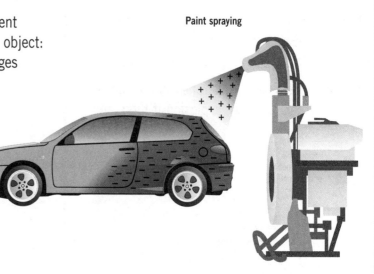

Paint spraying

Defibrillators

When the heart beats the heart muscle contracts. A **defibrillator** is used to start the heart when it has stopped.

The dotted lines show the path of the charge through chest, and heart:

- Two electrodes called paddles are placed on the patient's chest.
- The paddles must make a good electrical contact with the patient's chest.
- Everyone including the operator must 'stand clear' so they don't get an electric shock.

- The paddles are charged.
- The charge is passed from one paddle, through the chest to the other paddle to make the heart muscle contract.

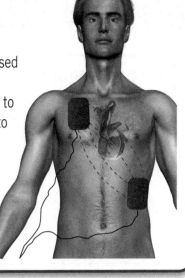

Crop Spraying

Fertiliser and insecticide spray nozzles are charged so that the droplets leaving the nozzle are charged. They repel each other and they are attracted to uncharged objects like the plants. The fine droplets cover the plant better and do not collect into large drops. They are less likely to drift in the wind and get wasted. This means that much less is used which saves money and is better for the environment.

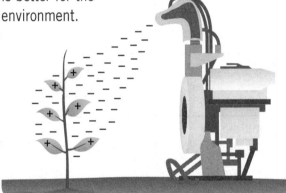

Build Your Understanding

The paddles take a few moments to charge up and then the discharge happens quickly. The electrons move through the heart muscle. Often the heart has not stopped, but has lost its steady rhythm. The defibrillator allows the heart to restart beating to its normal rhythm again.

⚠ Boost Your Memory

Remember these uses: Smoke, Sprays, and Shock.

✓ Maximise Your Marks

When you are describing how these applications work, explain what happens to the electrons in each case and how this affects the charge on the objects.

Electricity

❓ Test Yourself

1. Why do the plates in the electrostatic precipitator have the opposite charge to the grids?
2. What would happen if the paint drops had the same charge as the car body?
3. Give an advantage of electrostatic crop spraying.
4. Why is it important to make sure no one except the patient gets a shock from a defibrillator?

★ Stretch Yourself

1. Why are the grids in an electrostatic precipitator connected to a high voltage?
2. Sometimes metal objects are earthed instead of being given a charge. Explain how this works.

Electric Circuits

Circuit symbols

Component	Symbol	Component	Symbol
switch (open)		lamp	
switch (closed)		fuse	
cell		fixed resistor	
battery		variable resistor	
ammeter		light dependent resistor (LDR)	
voltmeter		thermistor	
junction of conductors		diode	
motor		generator	
power supply		a.c. power supply	

💡 Boost Your Memory

To learn the symbols, draw or print them onto cards – one card with the word and one card with the symbol. Then use them to play 'pairs'. Place them face down and turn two over at a time. If you get a 'pair' of the matching word and symbol you keep it. If not you place them face down again. The winner is the person with the most pairs.

✓ Maximise Your Marks

Take care when drawing circuit diagrams. Although the shape of the connecting wires does not matter they must join the components properly – electricity can't flow through gaps. The ammeter and voltmeter symbols are circles, not squares, and the symbol is 'A' not 'a'.

Electric Current

Electric current:
- Is a flow of **electric charge**.
- Only flows if there is a compete circuit. Any break in the circuit switches it off.
- Is measured in **amps** (A) using an **ammeter**.
- Is not used up in a circuit. If there is only one route around a circuit the current will be the same wherever it is measured.
- Transfers energy to the components in the circuit.

A **series circuit** is a circuit with only one route around it. The current measured on each ammeter will be the same.

A **parallel circuit** has more than one path for the current around the circuit. In this circuit there are two paths, marked in red and blue, around the circuit. The current measured on ammeters B and C adds up to give current measured on ammeter A and on ammeter D.

A series circuit

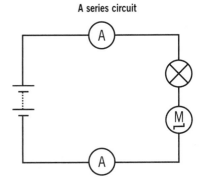

A parallel circuit

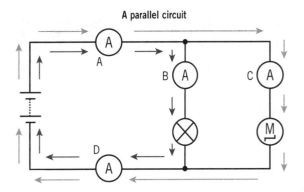

Build Your Understanding

Electric current is a flow of positive charge so the direction of the current is opposite to the direction of the electron flow, because electrons are negatively charged.

Electric charge is measured in coulombs (C). The amount of electric charge passing a point in the circuit depends on the current:

Charge (C) = current (A) × time (s)

$Q = I\,t$

Metal conductors contain lots of electrons that are free to move. When the battery makes the electrons move they flow in a continuous loop around the circuit. In insulators there are few charges that are free to move.

Batteries supply direct current, d.c. so the charges always move in the same direction from the positive terminal, around the circuit to the negative terminal. Mains electricity is produced by generators and the charges reverse direction. This is called alternating current (a.c.).

At a junction in a circuit, the total current flowing into the junction must be the same as the total current flowing out of the junction.

❓ Test Yourself

1. In a parallel circuit, if ammeter B reads 0.3 A and ammeter C reads 0.5 A what is the reading on:
   a) Ammeter A; b) Ammeter D?

2. In a parallel circuit, if ammeter B reads 500 mA and ammeter A reads 900 mA what is the reading on:
   a) Ammeter C; b) Ammeter D?

⭐ Stretch Yourself

1. If a current of 2 A is switched on for 10 s, how much charge has flowed?

2. In the parallel circuit above, what will always be true about the readings on ammeters A, B and C?

Voltage or Potential Difference

Voltage or Potential Difference

Voltage is also called **potential difference (p.d)**.

Voltage is:
- Measured between two points in a circuit.
- Measured in **volts** (V) using a **voltmeter**.

✓ Maximise Your Marks

Students often confuse voltmeters and ammeters. Always say **voltage across** and **current through**.

This will remind you that to measure the current flowing through a component you must connect the ammeter in line, so that the current flows through it. To measure the voltage across the component you must connect the voltmeter across the component making a connection on either side of it.

The higher the voltage of a battery the higher the 'push' on the charges in the circuit.

This diagram shows how to connect voltmeter A to measure the voltage supplied by the battery, and how to connect voltmeter B to measure the voltage across one of the lamps.

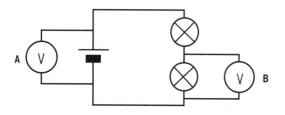

Build Your Understanding

Potential difference, or voltage is a measure of energy transferred to (or from) the charge moving between the two points.

In the diagram above:
- Voltmeter A is measuring the energy transferred *to* the charge.
- Voltmeter B is measuring the energy transferred *from* the charge.

The potential difference (voltage) between two points is the work done (energy transferred) per coulomb of charge that passes between the two points.

$$\text{Potential difference (V)} = \frac{\text{Work done (J)}}{\text{Charge (C)}}$$

$$V = \frac{W}{Q}$$

💡 Boost Your Memory

Remember that 'a volt is a joule per coulomb'. Add to this that 'a coulomb is an amp second' and you can work out most of the electricity relationships you need.

Voltage in Series or Parallel

When components are connected in **series** the voltage, or p.d., of the power supply is shared between the components.

Adding the measurements on the three voltmeters gives the power supply p.d. When components are connected in **parallel** to a power supply, the voltage, or p.d., across each component is the same as that of the power supply.

The measurements on all the voltmeters are the same.

Voltage in Series and Parallel Circuits

Adding the measurements on the three voltmeters gives the power supply p.d.

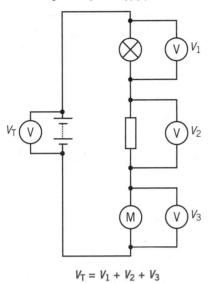

The measurements on all the voltmeters are the same

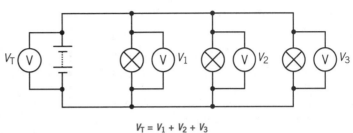

$V_T = V_1 + V_2 + V_3$

$V_T = V_1 + V_2 + V_3$

Build Your Understanding

Only identical cells should be connected together. In series the p.d. will be the sum of the p.d.s of the cells. In parallel, the p.d. is unchanged. The current will be larger when the cells are in series. In parallel, the current is unchanged, but the cells will last longer because there is more stored charge.

Cells connected in series and in parallel

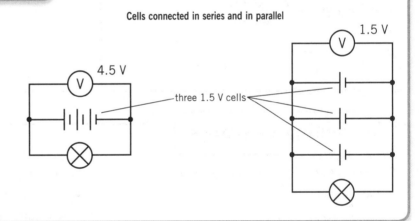

? Test Yourself

1 a) In the series circuit the battery voltage is 12 V, the voltage across the motor is 6 V and across the resistor is 4 V. What is the voltage across the lamp?

b) Does this mean that the current will be different in each component? Explain your answer.

2 a) In the parallel circuit above if the battery voltage is 9 V what is the voltage across each of the lamps?

b) Does this mean that the lamps will be equally bright? Explain your answer.

★ Stretch Yourself

1 If the voltage across a lamp is 9 V and 10 C of charge flows through the lamp how much energy has been transferred?

2 What voltage would be supplied by five 1.5 V cells:

a) In series.

b) In parallel.

3 What advantage is there to connecting the five cells in parallel?

Electricity

Resistance and Resistors

Resistance

The components and wires in a circuit **resist** the flow of electric charge. When the **voltage**, (or p.d.), *V*, is fixed, the larger the **resistance** of a circuit the less **current**, *I*, passes through it.

The resistance of the connecting wires is so small it can usually be ignored.

Other metals have a larger resistance, for example the filament of a light bulb has a very large resistance. Metals get hot when charge flows through them. The larger the resistance the hotter they get. A light bulb filament gets so hot that it glows.

Resistance is measured in **ohms** (Ω).

$$\text{Resistance } (\Omega) = \frac{\text{voltage (V)}}{\text{current (A)}}$$

$$R = \frac{V}{I}$$

Build Your Understanding

Metals are made of a lattice of stationary positive ions surrounded by free electrons. The moving electrons form the current. In metals with low resistance the electrons require less of a 'push' (p.d) to get through the lattice. The moving electrons collide with the stationary ions and make them vibrate more. This increase in kinetic energy of the lattice increases the temperature of the metal.

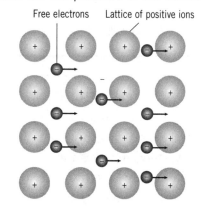

Free electrons   Lattice of positive ions

Combining Resistors

Components can be added to a circuit in **series** or in **parallel**.

For components in series:
- Two (or more) components in series have more resistance than one on its own.
- The current is the same through each component.
- The p.d. is largest across the component with the largest resistance.
- The p.d.s across the components add up to give the p.d. of the power supply.
- The resistances of all the components add up to give the total resistance of the circuit.

For components in parallel:
- Two (or more) components in parallel have less resistance than one on its own.
- The current through each component is the same is if it were the only component.
- The total current will be the sum of the currents through all the components.
- The p.d. across all the components will be the same as the power supply p.d.
- The current is largest through the component with the smallest resistance.

Resistors connected in series

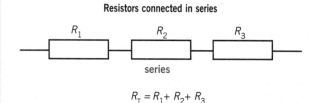

series

$$R_T = R_1 + R_2 + R_3$$

Resistors connected in parallel

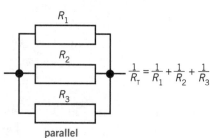

$$\frac{1}{R_T} = \frac{1}{R_1} + \frac{1}{R_2} + \frac{1}{R_3}$$

parallel

Build Your Understanding

For components in series:
- Two (or more) components in series have more resistance than one on its own. This is because the battery has to push charges through both of them.
- The p.d. is largest across the component with the largest resistance. This is because more work is done by the charge passing through a large resistance than through a small one.

For components in parallel:
- A combination of two (or more) components in parallel has less resistance than one component on its own. This is because there is more than one path for charges to flow through.
- The current is largest through the component with the smallest resistance. This is because the same battery voltage makes a larger current flow through a small resistance than through a large one.

⚠ Boost Your Memory

A series is one after the other, parallel lines are side by side, so series circuits have one component after another and parallel circuits have components that can be drawn side by side.

Fixed Resistors

In some components, such as **resistors** and **metal conductors**, the resistance stays constant when the current and voltage change, providing that the temperature does not change.

For this type of fixed resistance if the voltage is increased the current increases so that a graph of current against voltage is a straight line. The current is **directly proportional** to the voltage – doubling the voltage doubles the current. Components that obey this law (**Ohm's Law**) are sometimes called **ohmic** components.

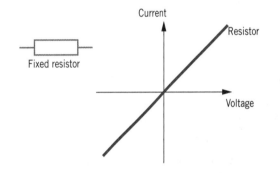

A graph of current against voltage for a resistor

Fixed resistor

✓ Maximise Your Marks

When there is no voltage there is no current, so graphs of I against V pass through the point (0,0). Remember this when you are drawing graphs.

? Test Yourself

1. Lamp voltage = 9 V current = 0.1 A. What is the resistance?

2. Voltage = 12 V resistance = 200 Ω. What is the current?

3. In the series circuit on page 80 R_1 =100 Ω, R_2 = 200 Ω, R_3 =300 Ω.
   a) What is the total resistance?
   b) Which resistor will have the largest voltage across it?

★ Stretch Yourself

1. Why do metals get hot when an electric current flows through them?

2. A student takes these measurements for a resistor:
   V = 3 V; I = 20 mA
   V = 4.5 V; I = 30 mA

   Explain whether the resistor obeys Ohm's Law.

Special Resistors

A Filament Lamp

The wire in a **filament lamp** gets hotter for larger currents. This increases the resistance so the graph of current against voltage is not a straight line.

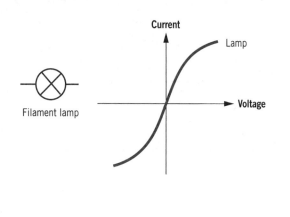

Variable Resistors

A **variable resistor** changes the current in a circuit by changing the resistance. This can be used to change how circuits work, for example to change:

- How long the shutter is open on a digital camera.
- The loudness of the sound from a radio loud speaker.
- The brightness of a bulb.
- The speed of a motor.

Inside one type of variable resistor is a long piece of wire made of metal with a large resistance (called **resistance wire**). To alter the resistance of the circuit a sliding contact is moved along the wire to change the length of wire in the circuit.

Special Resistors

The resistance of a **light dependent resistor (LDR)** decreases as the amount of light falling on it increases. It can be used in a circuit to switch a lamp on, or off, when it gets darker, or lighter.

The resistance of the most common type of **thermistor** (a negative temperature coefficient (NTC) thermistor) decreases as the temperature increases. It can be used in a circuit to switch a heater or cooling fan, on, or off, at a certain temperature.

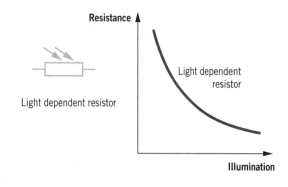

Current will only flow through a **diode** in one direction – the forward direction. In one direction its resistance is very low, but in the other direction, called the **reverse direction**, its resistance is very high.

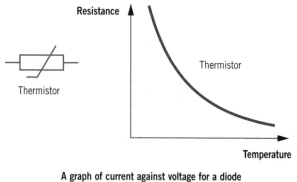

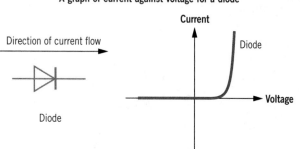

Build Your Understanding

Two resistors can be used in a circuit to provide an output p.d. with the value that is wanted from a higher input p.d. This is called a potential divider circuit.

current: $I = \dfrac{V_{in}}{(R_1 + R_2)} = \dfrac{V_1}{R_1} = \dfrac{V_2}{R_2}$

voltage: $V_{in} = V_1 + V_2$

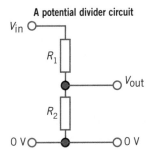

A potential divider circuit

The p.d.s (or voltages) are divided in the same ratio as the resistances.

When a thermistor is used as one of the resistors the resistance will change with the temperature. This circuit will produce a p.d. that changes with temperature and so it can be used to switch a heater on or off.

An LDR can be used in the same way.

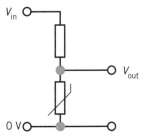

The temperature dependent potential divider

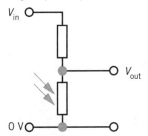

A light dependent potential divider

✓ Maximise Your Marks

When an ordinary resistor gets hotter its resistance increases, but for most common thermistors resistance decreases. When light intensity increases LDRs resistance decreases. The extra energy makes it easier for current to flow in these materials.

Light Emitting Diodes

A **light emitting diode (LED)** is a diode that emits light. LEDs are becoming widely used as low voltage and low energy sources of light.

Notice that symbols for diodes and LEDs (and also LDRs) sometimes include a circle:

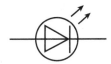

⚛ Boost Your Memory

The diode symbols are like arrow heads which show which way the current goes.

❓ Test Yourself

1. Why is the resistance of a filament lamp higher when it is switched on?

2. Why would the metal used to make lamp filaments be unsuitable for connecting components in a circuit?

3. What property of a thermistor makes it useful for controlling a heater?

4. Mains electricity is a.c. and computers require a d.c. supply. Which component would be useful in a circuit to connect a computer to use mains electricity? Why?

★ Stretch Yourself

1. In the potential divider circuit if the input p.d. $V_{in} = 5$ V, $R_1 = 300\ \Omega$, $R_2 = 200\ \Omega$. What is:
   a) V_1; b) V_2?

2. In the circuit with the thermistor when the temperature increases what happens to:
   a) The resistance of the thermistor?
   b) The p.d. across the thermistor?
   c) The p.d. across the resistor?

Electricity

The Mains Supply

Safe Use of Mains Electricity

- Mains voltage is 230 V a.c.
- The direction of the current and voltage changes with frequency = 50 Hz
- An electric shock from the mains can kill.

You should never interfere with mains electricity in any way. It should only be a professional electrician. The colour code for mains electricity cables used in buildings and appliances is shown below.

Name of wire	Colour of insulation	Function of the wire
Live	Brown	Carries the high voltage.
Neutral	Blue	The second wire to complete the circuit.
Earth	Green and yellow	A safety wire to stop the appliance becoming live.

This diagram shows the wiring of a 3-pin plug for a heater with a metal case. The **fuse** is always connected to the brown, **live** wire. A **cable grip** is tightened where the cable enters the plug to stop the wires being pulled out.

A **fuse** is a piece of wire that is thinner than the other wires in the circuit. It will melt first if too much current flows before the wires overheat.

A 3 A fuse will melt if a current of 3 A flows through it. Choose a fuse that is the lowest value, which is more than the normal operating current. If there is a fault, or if too many appliances are plugged into one socket, resulting in a large current, then the fuse will melt and break the circuit preventing a fire.

The **earth wire** is connected to the metal case of appliances so that when they are plugged into the mains supply the metal case is earthed (see page 73). If there is a fault and the live wire touches the metal case, a very large current flows through the low-resistance path to earth melting the fuse wire and breaking the circuit.

Double insulated appliances have cases that do not conduct (usually plastic) and have no metal parts that you can touch, so they do not need an earth wire.

A 3-in plug on an earthed appliance

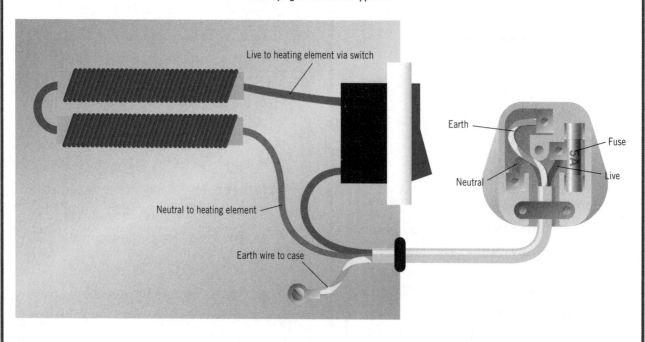

Live to heating element via switch

Earth

Fuse

Neutral

Live

Neutral to heating element

Earth wire to case

Electricity

Build Your Understanding

The fuse takes a short time to melt. It will not prevent you from getting an electric shock if you touch a live appliance. A residual current circuit breaker (RCCB) is safer.

These are switches to cut off the electricity very quickly if they detect a difference in the current flowing in the live and the neutral wires, (for example, if the current flows through a person, or appliance casing). Another advantage is that they can be switched back on once the fault is fixed, whereas a fuse must be replaced.

An RCCB can be part of mains circuit in a building, or a plug in device that goes between the appliance and the socket.

Appliances that are dangerous include:
- Those where the cable could get wet, or be cut, for example, lawn mowers and power tools.
- Music amplifiers connected to a metal instrument that someone is playing.

○ Boost Your Memory

Try explaining these safety features to someone. You'll soon find out if you remember them.

Electrical Power

The power is the rate at which the power supply transfers electrical energy to the appliance. It is measured in watts (W).

$$\text{power (W)} = \frac{\text{energy (J)}}{\text{time (s)}}$$

Electrical power (W) = current (A) × voltage (V)

$P = IV$

Electrical energy (J) = current (A) × voltage (V) × time (s)

$E = IVt$

Example: What is the current in a 2.8 kW kettle?

Using $P = I\,V$

$I = P \div V$

$I = 2800 \text{ W} \div 230 \text{ V} = 12.2 \text{ A}$

Power and Resistance

Another useful equation is:

Using $P = IV$ and $R = \dfrac{V}{I}$ so $V = IR$

$P = I \times (IR) = I^2R$

Power (W) = [current(A)]² × resistance (Ω)

$P = I^2R$

✓ Maximise Your Marks

Check carefully which equations you are given in the exam and which you need to learn. Make sure you know where to find them on the exam paper. You may find the triangle method useful for rearranging equations (see page 56).

? Test Yourself

1. Sam replaces a fuse with a piece of high resistance wire. Why is this a bad idea?

2. The earth wire is not connected to a metal appliance. Why is this dangerous?

3. What is the current in:
   a) A 2.5 kW kettle.
   b) A 9 W lamp.
   c) A 300 W TV.

4. Fuses come in 3 A, 5 A and 13 A. Which would you use for each appliance in question 3?

★ Stretch Yourself

1. Give two advantages of using an RCCB with outdoor Christmas lights.

2. Cables have 100 Ω resistance. Calculate the power wasted heating the cables when the current is:
   a) 0.5 A
   b) 1 A

Practice Questions

 Complete these exam-style questions to test your understanding. Check your answers on page 124. You may wish to answer these questions on a separate piece of paper.

1 A Perspex rod is rubbed with a duster and suspended from a piece of thread.

a) What happens when a second rubbed Perspex rod is brought close to it? Explain why. (2)

b) Rods of different plastics are brought close to the Perspex rod. The Perspex is attracted to rod A and repelled from rod B. What does this tell you about i) rod A and ii) rod B? (2)

2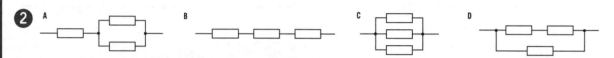

All the resistors in these combinations are identical.

a) Which of the combinations has the largest resistance? (1)

b) Which of the combinations has the smallest resistance? (1)

3 a) What type of meter is X? (1)

b) What type of meter is Y? (1)

c) Complete this table: (1)

	Current (mA)	Voltage (V)
Lamp	50	6
Resistor		

4 Refer to the diagram opposite. Calculate:

a) The current through resistor C. (1)

b) The voltage across resistor C. (1)

c) The voltage across lamp B. (1)

d) The current through resistor A. (1)

e) The resistance of A. (1)

5 Fill in the gaps to complete the sentences. Words can be used once, more than once or not at all. Choose from:

10 Ω   500 Ω   1000 Ω   0 V   0.15 V
5 V   10 V   15V

dark   LDR   LED   light
resistor   thermistor   variable

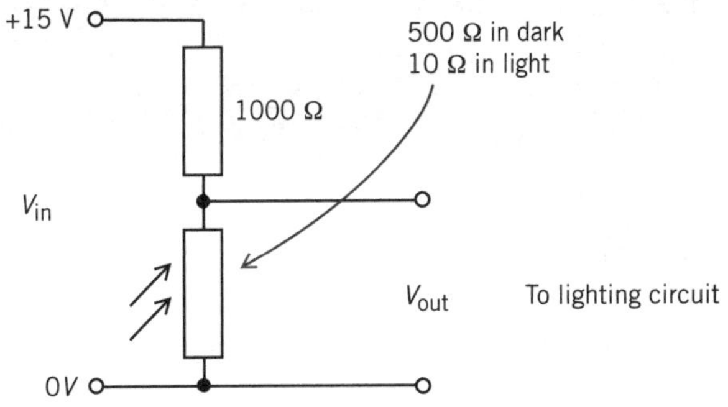

+15 V o———

1000 Ω

500 Ω in dark
10 Ω in light

V_{in}

V_{out} To lighting circuit

0V o———

When it is dark the resistance of the LDR is ....**a**........ and V_{out} = ...**b**......... When it is light the resistance of the LDR is ....**c**........ and ohms V_{out} = ...**d**........ This circuit can be used to switch a light on when it gets ....**e**......... The light level at which the light switches on or off can be altered by replacing the ....**f**........ ....**g**........ with a ....**h**........ ....**i**........ (2)

6 What size fuse should be fitted to a plug for a 500 W hairdryer?

A 1 A

B 3 A

C 5 A

D 13 A   (1)

7 An electric shower has a power of 8.0 kW when it is connected to the 230 V mains supply.

a) Calculate the current in the heater.   (1)

b) Explain why very thick cables are needed to connect it to the mains supply.   (1)

8 Tom works in a garage. He spray paints bare metal car parts using a method called electrostatic paint spraying. The spray gun is charged so that the paint droplets are all positively charged. Explain how this method works and the advantages of charging the paints droplets. (In your answer, aim to write between 8 and 12 lines.)   (6)

How well did you do?

| 0–7 | Try again | 8–15 | Getting there | 16–20 | Good work | 21–25 | Excellent! |

Atomic Structure

Radioactivity

The Atom

Atoms are about 10^{-10} m or 0.1 nanometres in diameter.

Neutral atoms have the same number of protons and electrons. When ionisation occurs atoms gain or lose electrons, becoming negatively or positively charged **ions**:

- The **atom** is mostly empty space with almost all the mass concentrated in the small positively charged nucleus at the centre.
- The nucleus is very small compared to the volume, or shell, around the nucleus that contains the **electrons**.

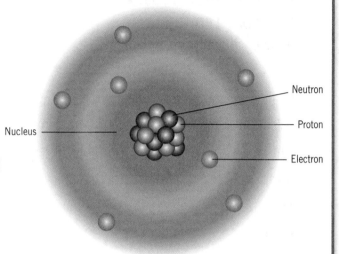

Nucleus

Neutron

Proton

Electron

Particles in the atom	Symbol	Where found in the atom	Relative mass	Relative charge
Proton	p	in nucleus	1	+1
Neutron	n	in nucleus	1	0 (neutral)
Electron	e	outside nucleus	$\frac{1}{1840}$	−1

The **nucleus** of an atom contains particles called **protons** and **neutrons**. These are also called **nucleons**.

The **atomic number** or **proton number**, **Z**, is the number of protons in the nucleus. The number of protons is what makes the atom into the element it is, so, for example, hydrogen always has one proton and carbon always has six.

The **mass number** or **nucleon number**, **A**, is the total number of protons and neutrons in the nucleus.

Isotopes of an element have the same number of protons in the nucleus, but different numbers of neutrons.

Isotopes of the same element have exactly the same chemical properties, but they have different mass and nuclear stability.

For example, carbon-12 is a stable isotope of carbon with 6 protons and 6 neutrons and carbon-14 is a radioactive isotope with 6 protons and 8 neutrons.

💡 Boost Your Memory

Make a set of flash cards with these words on one side and what they mean on the other. Keep looking at them and this will help you remember them.

Build Your Understanding

Nuclei are given symbols, for example this is the symbol for the stable isotope carbon-12 which has 6 protons and 6 neutrons: $^{12}_{6}\textbf{C}$

This is the symbol for an alpha particle (see page 90), which is the same as a helium nucleus. Sometimes α is used instead of He: $^{4}_{2}\textbf{He}$

A beta particle (see page 90) is not a nucleus, but this symbol is used for a beta particle in a nuclear equation. Sometimes β is used instead of e: $^{0}_{-1}\textbf{e}$

Nuclear Equations

Before and after a nuclear decay or reaction:
- The total of the mass numbers must be the same.
- The total of the atomic numbers must be the same.

Example: Alpha decay of radon-220.

$$^{220}_{86}\text{Rn} \rightarrow ^{216}_{84}\text{Po} + ^{4}_{2}\text{He}$$

$220 = 216 + 4$ and $86 = 84 + 2$

Example: Beta decay of carbon-14.

$$^{14}_{6}\text{C} \rightarrow ^{14}_{7}\text{N} + ^{0}_{-1}\text{e}$$

$14 = 14 + 0$ and $6 = 7 + (-1)$

A Model of the Atom

Before 1910 scientists had a **plum pudding** model of the **atom.**

The atom is made of positively charged material (pudding) with negatively charged electrons (plums) inside.

Ernest Rutherford investigated the structure of atoms by firing **alpha particles** at **gold foil**.

What happened to the alpha particles	Ernest Rutherford's explanation – the nuclear atom
Most went straight through the foil, without being deflected.	The atom is mostly empty space.
Some were deflected and there was a range deflection angles.	Parts of the atom have positive charge.
A very small number were 'back-scattered' – they came straight back towards the alpha particle source.	There is a tiny region of concentrated mass and positive charge which repels the very small number of alpha particles that have a head-on collision.

Build Your Understanding

The experiment was done in a vacuum, so that the alpha particles were not stopped by the air and it was surrounded by a fluorescent screen, so that a small flash of light was seen when an alpha particle hit the screen.

Hans Geiger and Ernest Marsden counted small flashes of light at different angles for hours.

Alpha particles scattering by gold foil

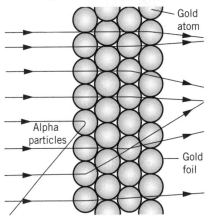

Gold atom

Alpha particles

Gold foil

✓ Maximise Your Marks

Remember: Most straight through, some deflected, very few straight back.

❓ Test Yourself

1. Which particles are found in the:
   a) atom; **b)** nucleus?

2. The atomic number of oxygen is 8. What does this tell you?

3. Which two are isotopes of an element: nitrogen-14, nitrogen-16, oxygen-16?

⭐ Stretch Yourself

1. Write an equation for:

   a) The alpha decay of uranium-235 (symbol= U, $Z = 92$) to thorium (symbol = Th).

   b) The beta decay of nitrogen-16 (symbol = N, $Z = 7$) to oxygen (symbol = O).

Radioactive Decay

Radioactive Emissions

There are three main types of radioactive emissions: **alpha particles**, **beta particles** and **gamma rays**.

Radiation	Ionising	Electric charge	Stopped by...	Affected by electric and magnetic fields?
Alpha (α)	Very strongly	–	A few cm of air. A sheet of paper.	Yes
Beta (β)	Yes	+	A thin sheet of aluminium.	Yes
Gamma (γ)	Only weakly	Neutral	A thick lead sheet. Thick concrete blocks.	No

When alpha and beta particles are emitted the nucleus changes into a different element. When gamma rays are emitted the element does not change.

Build Your Understanding

Alpha, Beta and Gamma

Alpha emission is when two protons and two neutrons leave the nucleus as one particle, called an alpha particle. It is identical to a helium nucleus.

Alpha particle (α)

Beta emission is when a neutron decays to a proton and an electron and the high energy electron leaves the nucleus as a beta particle.

Beta particle (β)

Gamma emission is when the nucleus emits a short burst of high-energy electromagnetic radiation. The gamma ray has a high frequency and a short wavelength.

Gamma rays (γ)

✔ Maximise Your Marks

Common mistakes are to say:
- 'An alpha particle is a helium *atom'*, but it is a helium *nucleus*.
- 'A beta particle is an electron from the *atom'*, but it is an electron emitted by a neutron in the nucleus when the neutron turns into a proton.

Radioactive Decay

A radioactive material contains nuclei that are unstable and emit nuclear radiation. This process is called **radioactive decay**. Radioactive decay is **random**. It is not possible to predict when it will happen, or make it happen by a chemical or physical process, for example by heating the material.

A radioactive source contains millions of atoms. The number of radioactive emissions a second depends on two things:
- The type of nucleus – some combinations of protons and neutrons are more stable than others.
- The number of undecayed nuclei in the sample – with double the number of nuclei, on average, there will be double the number of emissions per second.

Over a period of time the **activity** of a source gradually reduces.

The **half-life** of an isotope is the average time taken for half of the nuclei present to decay.

Radioactive Decay (cont.)

Example: Technetium-99m (Tc-99m) decays by gamma emission to Technetium-99 (Tc-99) with a half-life of six hours. After six hours, on average, only half of the Tc-99m nuclei remain. After another six hours, on average, only one quarter are left.

This pattern is the same for all isotopes, but the value of the half-life is different. Carbon-14 has a half-life of 5730 years; some isotopes have a half-life of less than a second.

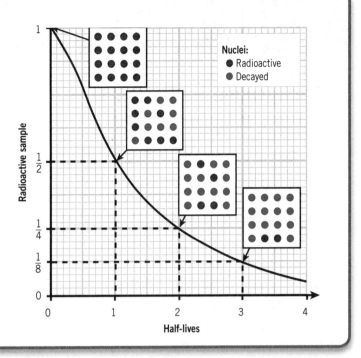

Build Your Understanding

After 10 half-lives, the activity has dropped to less than one thousandth of the original activity. This is often used as a measure of the time for a sample to decay to a negligible amount.

The number of radioactive emissions a second is called the activity of the source. The activity is measured in bequerel (Bq). An activity of 1 Bq is one radioactive emission per second.

After this number of half-lives...	The activity has dropped to this fraction of the initial value...	Which is the same as...
1	$\frac{1}{2}$	$\frac{1}{2}^{1}$
2	$\frac{1}{2} \times \frac{1}{2} = \frac{1}{4}$	$\frac{1}{2}^{2}$
3	$\frac{1}{2} \times \frac{1}{2} \times \frac{1}{2} = \frac{1}{8}$	$\frac{1}{2}^{3}$
10	$\frac{1}{1024}$	$\frac{1}{2}^{10}$

✓ Maximise Your Marks

Common mistakes are to say:
- After three half-lives there are $\frac{1}{3}$ or $\frac{1}{6}$ of the radioactive nuclei left – but it is $\frac{1}{8}$.
- $\frac{1}{8}$ of the nuclei '*have decayed*' – but it is $\frac{7}{8}$ because only $\frac{1}{8}$ are left.

? Test Yourself

1. Answer alpha, beta or gamma.

   a) Which has/have a positive charge?

   b) Which is/are stopped by a thin sheet of aluminium?

2. A radioactive source has a half-life of 24 hours. What fraction will remain after one day and four days?

★ Stretch Yourself

1. An alpha particle loses energy and attracts an electron. What has it become?

2. A radioactive source with a half-life of three hours has an activity of 16000 Bq. What is the activity after:
   a) 3 hours; b) 9 hours; c) 30 hours;
   d) When will the activity be 500 Bq?

Living with Radioactivity

Background Radiation

Radioactive materials occur naturally or are man-made. **Cosmic rays** from space make some of the carbon dioxide in the atmosphere radioactive. The carbon dioxide is used by plants and enters food chains. This makes all living things radioactive.

Some rocks are radioactive. We receive a low level of radiation from these sources all the time. It is called **background radiation**. Background radiation comes from:

- Radon (a radioactive gas) from rocks.
- Soil and building materials.
- Medical and industrial uses of radioactive materials.
- Food and drink.
- Cosmic rays (from outer space).
- 'Leaks' from radioactive waste and nuclear power stations.

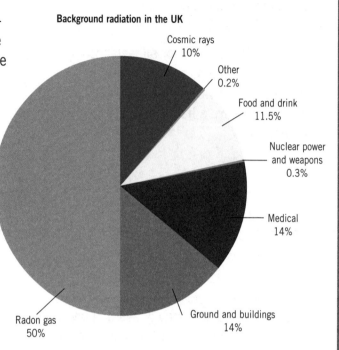

Background radiation in the UK

Cosmic rays 10%
Other 0.2%
Food and drink 11.5%
Nuclear power and weapons 0.3%
Medical 14%
Ground and buildings 14%
Radon gas 50%

Build Your Understanding

Some rocks are more radioactive than others, so the level of background radiation depends on the underlying rocks. Radon from some rocks builds up in houses. It emits alpha radiation, so it is particularly damaging in the lungs. Houses with high levels of radon can have under-floor fans fitted to keep the radon out of the house.

In some parts of the UK the rocks are more radioactive than in others and there is a higher level of background radiation.

✓ Maximise Your Marks

Background radiation is the name given to radioactive emissions from nuclei in our surroundings. Do not confuse this with radiation from mobile phones or cosmic microwave background radiation.

Sources of background radiation: cosmic rays are from outer space (not the Sun).

Food and drink are radioactive as explained above – not because of food irradiation.

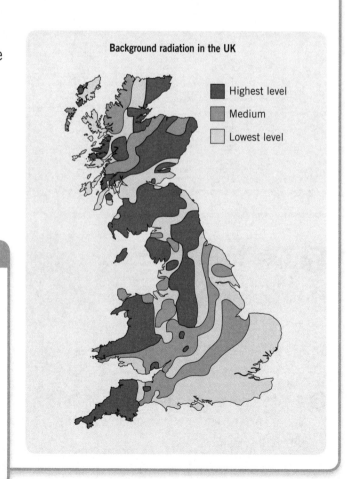

Background radiation in the UK

- Highest level
- Medium
- Lowest level

Dangers of Radiation

Ionising radiation is radiation that has enough energy to charge and ionise the atoms in molecules (see page 88/89). The ions can take part in chemical reactions. In living cells this can damage or kill them. It can damage the DNA so that the cell mutates into a cancer cell.

It is not possible to predict which cells will be damaged by exposure to radiation or who will get cancer.

Scientists studied the survivors of incidents where people were exposed to ionising radiation. They measured the amount of exposure and recorded how many people later suffered from cancer. The risk of cancer increases with increased exposure to radiation. People tend to overestimate the risk from radiation because it is invisible and unfamiliar. They underestimate the risk of familiar activities, like smoking.

Contamination and Irradiation

There are two types of danger from radioactive materials:
- **Irradiation** is exposure to radiation from a source outside the body.
- **Contamination** is swallowing, breathing in, or getting radioactive material on your skin.

A short period of irradiation is not as dangerous as being contaminated because, once contaminated, a person is continually being irradiated.

Alpha radiation has very short range. Even if it reaches the skin it does not penetrate, so there is little danger from irradiation. But it is strongly ionising, so contamination by an alpha source, for example breathing in radon gas, can be very

dangerous. Once inside the lungs it will keep irradiating sensitive cells.

Beta radiation has longer range and penetrates the skin so there is more danger from irradiation. It is not as strongly ionising as alpha, so contamination is not as dangerous as with alpha radiation.

Gamma radiation is long range and passes right through the body. There is a danger from irradiation if the radiation levels are high. It is only weakly ionising and most of the rays pass right through the body without hitting anything so contamination is less dangerous than with alpha or beta radiation.

Build Your Understanding

Radiation dose, measured in sieverts (Sv), is a measure of the possible harm done to the body. Radiation dose depends on the type of radiation, the time of exposure, and how sensitive the tissue exposed is to radiation. The dose is linked to the risk of cancer developing. Alpha radiation is strongly ionising and so the dose is 20 times larger from alpha than from beta or gamma radiation.

The normal UK background dose is a few milliSieverts. A fatal dose is between 4 Sv and 5 Sv given in one go.

? Test Yourself

1. How much of the background radiation in the UK is from radon gas?

2. Give one effect of ionising radiation on living cells.

3. 'She has been irradiated, she will get cancer'. What is wrong with this statement?

★ Stretch Yourself

1. Name a part of the UK that has high background radiation.

Uses of Radioactive Materials

Choosing the Best Isotope

For each use of radioactive materials:
- Alpha, beta or gamma radiation is chosen depending on the **range** and the **absorption** required.

An isotope is chosen depending on:
- whether it emits alpha, beta or gamma radiation
- how long it remains radioactive, which depends on the half-life.

Build Your Understanding

When radioactive materials are used, we have to decide whether the benefits outweigh the risks.

Example: Treatment benefits patients, but does not benefit hospital staff who work with radioactive materials regularly.

Radiation workers have their exposure monitored and take safety precautions to keep their dose as low as possible:
- They wear protective clothing.
- They keep a long distance away, for example they use tongs to handle sources.
- They keep their exposure time short.
- Sources should be shielded and labelled with the radioactive symbol.

💡 Boost Your Memory

Use a mnemonic that contains the first letters of each to remember safety precautions: **M**y **c**urtains **d**im **t**he **b**right **l**ight

Monitor, **c**lothing, **d**istance, **t**ime, **b**arrier, **l**abel.

Medical and Health Uses

Medical Tracers

Isotopes which emit **gamma radiation** (or sometimes **beta radiation**) are used in **medical tracers**. The patient drinks, inhales, or is injected with the tracer which is chosen to collect in the organ doctors want to examine.

Example: Radioactive iodine is taken up by the thyroid gland, which can then be viewed using a **gamma camera** that detects the gamma radiation passing out of the body. The tracer must not decay before it has moved to the organ being investigated, but it must not last so long that the patient stays radioactive for weeks afterwards.

Treating cancer

Isotopes which emit a higher dose of **gamma radiation** than tracers are used to build up in the cancer and kill the cancer cells.

Alternatively, beams of **gamma rays** are concentrated on a tumour to kill the cancer cells.

Example: Cobalt-60 emits high energy gamma rays and remains radioactive for years.

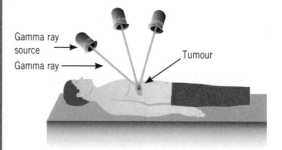

Gamma ray source
Gamma ray
Tumour

Sterilisation

Gamma radiation destroys microbes and is used for:
- Sterilising equipment, for example surgical instruments.
- Food irradiation to extend the shelf-life of perishable food.

The food or equipment does not become radioactive because it is only **irradiated**, there is no **contamination**. It is also irradiated inside the plastic packaging, so that it stays sterile.

More Uses of Radioactive Isotopes

Smoke detectors

Isotopes which emit **alpha radiation** are used in **smoke detectors**. The alpha radiation crosses a small gap and is picked up by a detector. If smoke is present, the alpha radiation is stopped by the smoke particles. No radiation reaches the detector and the alarm sounds. Beta and gamma radiation are unsuitable because they pass through the smoke.

Tracers

Isotopes which emit beta radiation or gamma radiation are used as tracers. Because a tracer is radioactive, detectors can track where it goes. A tracer is added to sewage as it enters a river, to trace its movement. The isotope used has an activity that will fall to zero quickly after the test is done.

Dating

The amount of radioactive carbon left in old materials that were once living can be used to calculate their age.

Some rocks contain a radioactive isotope of uranium that decays to lead, so they can be dated by comparing the amounts of uranium and lead. The more lead there is the older the rock is.

Build Your Understanding

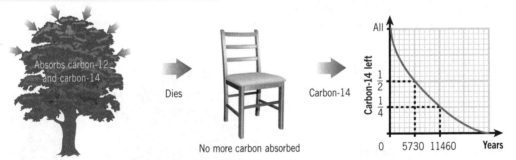

Absorbs carbon-12 and carbon-14

Dies

No more carbon absorbed

Carbon-14

Carbon dating:
- Can only be used for things that once lived.
- Cannot be used for objects older than 10 half-lives = 10 × 5730 years or less than 100 years.

✓ Maximise Your Marks

Learn the properties of alpha, beta and gamma radiation, so that you can explain why each is chosen for different areas.

❓ Test Yourself

1 Give a use of:
  a) Alpha radiation.
  b) Beta radiation.
  c) Gamma radiation.

2 Explain why healthy tissue is not killed by the gamma rays from the cobalt-60 during cancer treatment.

★ Stretch Yourself

1 Why is alpha radiation used in smoke detectors?

2 Why is carbon dating not used to confirm the age of a 60 000 year old egg?

Nuclear Fission and Fusion

Nuclear Fission

Nuclear fission is when a nucleus splits into two nuclei of about equal size, and two or three neutrons.

Example: After uranium-235 absorbs a neutron:

A neutron absorbed by a uranium nucleus causes a nuclear fission

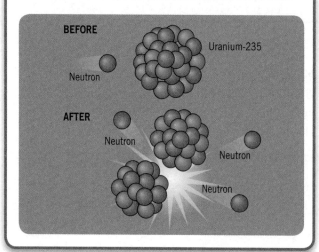

BEFORE

Uranium-235

Neutron

AFTER

Neutron

Neutron

Neutron

Build Your Understanding

A small amount of mass is converted into a large amount of energy. The energy is calculated using Einstein's equation:

Energy (J) = mass (kg) [speed of light in a vacuum (m/s)]²

$$E = mc^2$$

About a million times more energy is released than in a chemical reaction.

A Chain Reaction

After the new neutrons slow down, they can strike more uranium nuclei and cause more fission events. This is called a **chain reaction**.

If the chain reaction runs out of control it is an **atomic bomb**. In a **nuclear reactor** the process is controlled. In a nuclear power station the energy released is used to generate electricity.

Nuclear Power

In nuclear reactors:
- The **fuel rods** are made of **uranium-235** or **plutonium-239**.
- The **moderator** is a material that slows down the neutrons, so they can be absorbed.
- The energy heats up the reactor core.
- A **coolant** is circulated to remove the heat.
- The hot coolant is used to heat water to steam, to turn the power station turbines.
- The **control rods** are moved into the reactor to absorb neutrons to slow or stop the reaction.

Reactors produce **radioactive waste**. The half-lives of some isotopes are thousands of years, so radioactive waste must be kept safely contained for thousands of years.

A nuclear reactor

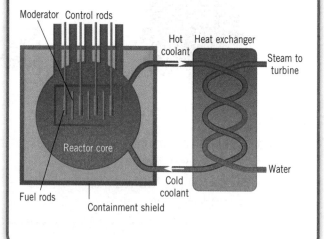

Moderator Control rods

Hot coolant   Heat exchanger

Steam to turbine

Reactor core

Water

Cold coolant

Fuel rods

Containment shield

A chain reaction

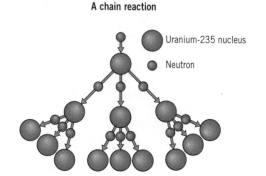

Uranium-235 nucleus

Neutron

Radioactivity

Radioactive Waste

Types of radioactive waste	Examples	Disposal
Low level waste	Used protective clothing	Sealed into containers. Put into landfill sites.
Intermediate level waste	Material from reactors	Mixed with concrete. Stored in stainless-steel containers.
High level waste	Used fuel rods	Kept under water in cooling tanks (it decays so fast it gets hot). Eventually becomes intermediate level waste.

More About Fusion

The protons and neutrons inside the nucleus are held together by a force called the **strong force**.

The problem is getting, and keeping, the temperature and pressure high enough to overcome the repulsive force, so that the nuclei get close enough for the strong force to take over. When fusion happens a small amount of mass is converted into a large amount of energy ($E=mc^2$).

🔅 Boost Your Memory

Draw up a table to compare fusion and fission. It will help you when you revise.

✓ Maximise Your Marks

You may be asked for advantages and disadvantages, for example, of methods of waste disposal or nuclear reactors. To gain full marks you must give at least one advantage *and* one disadvantage.

If you are asked for your opinion it doesn't matter if you say 'yes' or 'no', but you must say one or the other and give a reason. Remember that most of the marks are for justifying your choice.

Build Your Understanding

Where is the safest place to store the radioactive waste?

- At the bottom of the sea, but the containers may leak.
- Underground, but the containers may leak and earthquakes or other changes to the rocks may occur.
- On the surface, but needs guarding from terrorists, for thousands of years.
- Blast into space, but there is a danger of rocket explosion.

Nuclear Fusion

When two small nuclei are close enough together they can **fuse** together to form a larger nucleus. This releases a large amount of energy. The problem is getting the two nuclei close enough, because nuclei are positively charged and **repel** each other. Inside stars the temperatures are high enough for the nuclei to have enough energy for nuclear fusion to occur.

Scientists are trying to control the reaction and design nuclear fusion reactors.

❓ Test Yourself

1. What is the difference between nuclear fission and nuclear fusion?
2. Describe a chain reaction in plutonium-239.
3. Explain what control rods are used for.
4. How are fuel rods disposed of when the fuel is used up?

★ Stretch Yourself

1. How much energy results from 0.1 g of fuel being converted to energy?
2. What is the force that holds protons and neutrons together?

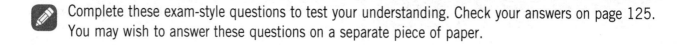

Practice Questions

Complete these exam-style questions to test your understanding. Check your answers on page 125. You may wish to answer these questions on a separate piece of paper.

1 Nuclear radiation passes through a thin aluminium sheet. The radiation could be:

A Alpha radiation only.

B Beta radiation only.

C Alpha, beta or gamma radiation.

D Beta or gamma radiation. (1)

2 A patient is given a dose of iodine-131 which has a half-life of eight days. What fraction of the iodine-131 nuclei are left after 32 days?

.... (1)

3 $^{235}_{92}$U and $^{238}_{92}$U are both isotopes of uranium. Which of these tables is correct: (1)

A

Isotope	Protons	Neutrons
$^{235}_{92}$U	92	235
$^{238}_{92}$U	92	238

B

Isotope	Protons	Neutrons
$^{235}_{92}$U	92	143
$^{238}_{92}$U	92	146

C

Isotope	Protons	Neutrons
$^{235}_{92}$U	235	92
$^{238}_{92}$U	238	92

D

Isotope	Protons	Neutrons
$^{235}_{92}$U	143	92
$^{238}_{92}$U	146	92

4 A patient is injected with Technetium-99 m which has a half-life of six hours.

a) What fraction of the Technetium-99m nuclei are left after one day? .... (1)

b) Why is it better to use an isotope with a half-life of six hours rather than:

i) An isotope with a half-life of six minutes .... (1)

ii) An isotope with a half-life of six days. .... (1)

5 Answer the questions below.

a) Describe how the alpha particle scattering experiment was done by Hans Geiger and Ernest Marsden.

.... (3)

b) What three effects did they observe?

.... (3)

c) What did Rutherford conclude from these results?

.... (2)

6 Answer the questions below.

a) Sodium-24 is a radioactive isotope of sodium that emits gamma radiation.
The stable form of sodium is sodium-23. Explain how the nuclear structure of
sodium-24 is different to sodium-23. (2)

....

The activity of a sodium-24 source was measured over two days and this graph was plotted.

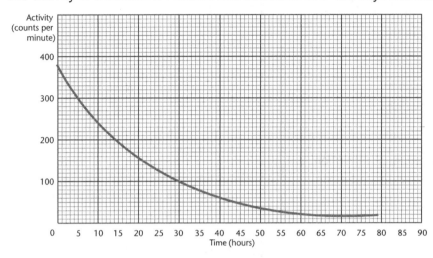

b) Use the graph to work out the half-life of sodium-24. (2)

....

c) Use the graph to estimate the background activity at the site of the experiment.

....

d) Explain whether a salt of sodium-24 would be a suitable isotope to use for:

i) The source in a smoke detector. (2)

....

ii) A tracer to find the leak in an underground water pipe.

....

7 Anna has a tumour of the thyroid gland. The thyroid gland absorbs iodine. Anna can be cured
with an injection of radioactive iodine-131. This is a beta emitter with a half-life of 8 days.
Write a patient leaflet for patients like Anna explaining how the treatment works and what
she should take into account when deciding whether to have the treatment. (In your answer,
aim to write between about 8 and 12 lines.) (6)

....

....

....

How well did you do?

| 0–7 | Try again | 8–16 | Getting there | 17–21 | Good work | 22–26 | Excellent! |

Refraction, Dispersion and TIR

Refractive Index

Refraction occurs when light changes speed (see pages 20–21). The wavelength changes, but the frequency of the light stays the same.

The **refractive index**, *n*, of a medium:

$$n = \frac{\text{speed of light in a vacuum}}{\text{speed of light in the medium}}$$

The refractive index is a ratio, so it has no units.

Snell's law states that the **angle of refraction**, *r*, when light enters a medium depends on the angle of incidence, *i* and the refractive index:

$$n = \frac{\sin i}{\sin r}$$

Lenses use refraction to change the direction of light rays.

Refraction through a glass block

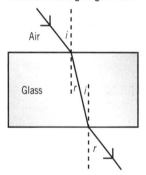

Calculating the Speed of Light

Example: Find the speed of light, *v* in a glass block.

Speed of light in a vacuum = 3×10^8 m/s. (The speed of light in air is so close to this that you can use this value, unless you are told otherwise.)

angle *i* = 30° angle *r* = 19°

$$n = \frac{\sin 30°}{\sin 19°} = 1.54$$

$$\frac{3 \times 10^8 \text{ m/s}}{v} = 1.54$$

$$v = 1.95 \times 10^8 \text{ m/s}$$

Dispersion

A prism separates white light into its different colours.

Dispersion happens because the speed of light in the glass depends on the colour (the frequency) of the light.

When white light passes from air to glass, violet rays are refracted towards the normal more than red rays.

When the light passes from glass to air the violet rays are refracted away from the normal more than red rays.

In a prism this results in the white light being spread into a spectrum, as each colour from red through to violet is refracted more than the last.

Dispersion by a glass prism

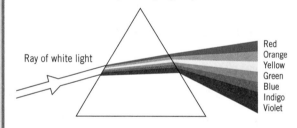

💡 Boost Your Memory

Remember which colour bends the most by the mnemonic: **V**iolet **V**eers **V**iolently.

Build Your Understanding

Dispersion happens because all electromagnetic waves travel at the same speed in a vacuum, but in other media the speed of light depends slightly on the frequency of the radiation.

Another way of saying this is that the refractive index depends on the frequency of the radiation.

The higher the frequency, the greater the change in speed and refractive index.

Total Internal Reflection

Total Internal Reflection (TIR) occurs when the angle of refraction is greater than 90° and the light cannot leave the medium (see page 35).

It only happens when light passes from a dense to a less dense medium, for example from glass to air, which means the angle of refraction is larger than the angle of incidence.

The **critical angle** is the angle of incidence above which TIR occurs. It is the angle of incidence that gives an angle of refraction of 90°.

The higher the refractive index, the lower the critical angle. Diamond has very high refractive index and so a low critical angle which leads to more internal reflections. This is why diamonds sparkle so much.

Some applications of TIR are below:

Reflecting prisms which are used in:
- Binoculars.
- Cats-eyes in the road to reflect headlights.
- Cycle reflectors.

Optical fibres are used in:
- Communications (see pages 34–35).
- Endoscopes.

An **endoscope** is two sets of **optical fibres** used for viewing inside the body or for performing minor surgical operations. Optical fibres are flexible so the endoscope can be passed into the body through an opening, for example the mouth or nose. The first set of fibres transmits light from an external light source into the body and the second set transmits the image out of the body.

Endoscopes can have tools attached for minor surgery. The optical fibres can also transmit a laser light beam, which can be used to cut and heat tissue to stop bleeding.

Total Internal Reflection in Prisms

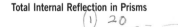

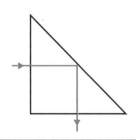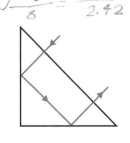

Build Your Understanding

The critical angle, *c*, of a medium with refractive index, *n*, is calculated using the equation:

$$\sin c = \frac{1}{n}$$

✓ Maximise Your Marks

Remember that TIR only happens when light is leaving the denser medium, for example from glass to air or from water to air.

❓ Test Yourself

1. Light is refracted away from the normal when it leaves water. What happens to its speed?

2. Which colour light travels slowest in glass?

3. Give one use of TIR.

4. Why does an endoscope need two bundles of optical fibres?

⭐ Stretch Yourself

1. When $i = 20°$ $r = 8°$ calculate the refractive index of diamond.

2. Calculate the critical angle for glass ($n = 1.5$).

Lenses

Converging Lenses

A **converging lens** makes a **parallel** beam of light rays converge to a point, called the **focus**, **focal point** or **principal focus**. All converging lenses are fatter in the middle than at the edge.

The **focal length**, *f*, of a lens is the distance from the centre of the lens to the focal point.

These lenses are also called **convex lenses**. When two lenses are made of the same material, the one with the most curved surfaces will have the shortest focal length.

Light from a source spreads out, or **diverges**. A converging lens can be used to reduce the divergence and make a parallel or converging beam.

Using a converging lens to make a parallel or converging beam

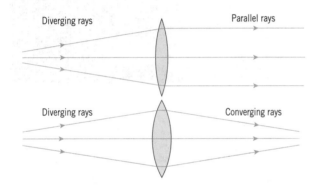

Diverging rays   Parallel rays

Diverging rays   Converging rays

Converging lenses with different focal lengths

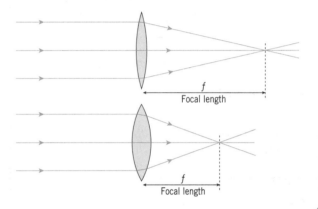

f
Focal length

f
Focal length

Build Your Understanding

Power is measured in dioptres (D) where 1 D = 1 m^{-1}. The shorter the focal length, the more powerful the lens.

Power (D) = $\dfrac{1}{\textbf{focal length (m)}}$

Examples:

- If a lens has a focal length of 0.8 m it has a power of 1 ÷ 0.8 = 1.25 D.
- A lens with a focal length of 10 cm has a power of 1 ÷ 0.1 = 10 D.
- Converging lenses have positive powers (for example +2 D) diverging lenses have negative powers (for example -0.5 D).

The magnification produced by a lens is:

$$\frac{\textbf{The image height}}{\textbf{The object height}}$$

Magnification has no units.

Images

A lens is used to produce an **image**. The image is a copy of the **object**.

Image	Meaning
Real	It can be displayed on a screen because the light passes through it.
Virtual	It can only be seen by looking through the lens – the light doesn't pass through it.
Upright	The same way up as the object.
Inverted	Upside down compared to the object.
Magnified	Bigger than the object.

A converging lens can form real, inverted images, or virtual, upright images depending how far it is from the object.

Build Your Understanding

You can find out what an image is like, and where it appears, by drawing a ray diagram using these facts:

- Light rays pass straight through the optical centre of the lens.
- Light rays parallel to the axis pass through the focal point (or light rays passing through the focal point leave the lens parallel to the axis).

What to do:

- Choose the scale.
- Draw a vertical line to represent the lens and a horizontal line to represent the principal axis

through the centre of the lens.
- Mark the focal point in the right place on each side of the lens.
- Draw the object – an upright arrow the right size and in the right place.
- Draw a ray from the top of the arrow straight through the optical centre of the lens.
- Draw a ray from the top of the arrow parallel to the axis to the lens, and then through the focal point.
- Draw the image arrow, from the point where the rays meet to the axis.

Ray Diagrams

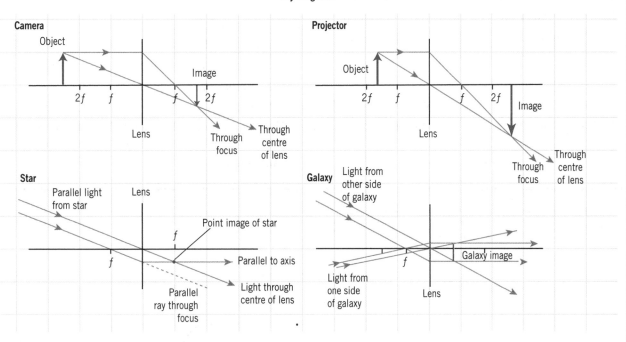

⚟ Boost Your Memory

Remember that a real image is one you can touch, a virtual image is not really there, like the image in a mirror or seen through a magnifying glass.

✓ Maximise Your Marks

Ray diagrams are best drawn on graph paper with a sharp pencil and a transparent ruler (so you can see the whole diagram).

❓ Test Yourself

1. Lens A has a focal length of 10 cm and lens B has a focal length of 15 cm. Which lens is the fatter?

2. A converging lens brings light to a focus, but the focal point is too far away from the lens. Should a fatter or thinner lens be used?

★ Stretch Yourself

1. What is the focal length of a lens with power = 4 D?

2. The object measures 2.4 cm and the image measures 6 cm. What is the magnification?

Seeing Images

The Eye

In the eye the convex lens and the cornea focus an image on the light sensitive cells of the **retina**. From these cells, signals are sent along the **optic nerve** to the brain.

The amount of light entering the eye depends on the size of the **pupil**, which is controlled by the **iris**.

The **cornea** is transparent and curved to refract light. The **ciliary muscles** alter the shape of the variable **lens**, so that the eye can focus on near and distant objects.

The average adult human eye can focus on objects between about 25 cm (the **near point**) and infinity from which rays are parallel (the **far point**).

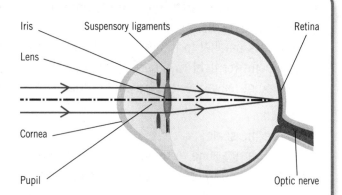

Spectacles, or contact lenses, are worn to correct:
- **Short-sight** where near objects are seen clearly, but distant objects are blurred. Corrected with a **diverging lens**.
- **Long-sight** where distant objects are seen clearly, but near objects are blurred, or viewing them causes eye strain. Corrected with a **converging lens**.

Build Your Understanding

Lenses for correcting eyesight

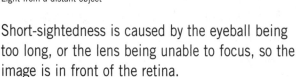

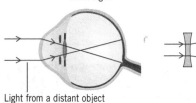
Diverging lens — Short sight — and its correction — Light from a distant object

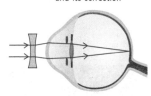

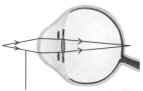
Converging lens — Long sight — and its correction — Light from a near object

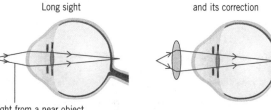

Short-sightedness is caused by the eyeball being too long, or the lens being unable to focus, so the image is in front of the retina.

Long-sightedness is caused by the eyeball being too short, or the lens being unable to focus, so the image is behind the retina. The focal length of the correcting lens depends on its refractive index and its curvature, so special glass and plastics with high refractive index are useful for making lenses which are thinner and flatter. Laser surgery is an alternative way of correcting eyesight. A laser is used to alter the curvature of the cornea so that the light rays are focused on the retina.

✓ Maximise Your Marks

If you are asked to compare the eye and the camera don't make statements only about the eye or only about the camera.

💡 Boost Your Memory

In short-*sight*, you *can see* things a short distance away. To stop the eye focusing in such a short distance needs a diverging lens.

In long-*sight*, you *can see* things a long way away. To stop the eye needing such a long distance to focus needs a converging lens.

Uses of Lenses

The converging lens in the camera produces a real, inverted image that is smaller than the object. The lens must be more than twice the focal length from the object so that it produces a small image, (see pages 102–103).

In many ways the camera is similar to the eye (see page 28). But to focus the image the camera lens is moved:
- Away from the film to focus on a close object.
- Towards the film to focus on a distant object.

The converging lens in the **projector** produces a real, inverted image that is magnified. The lens must be between one and two times the focal length from the slide or film so that it produces a magnified image (see page 103).

To focus the image the projector lens is moved:
- Away from the slide to focus on a closer screen and give a smaller image.
- Towards the slide to focus on a distant screen and give a more magnified image.

When a converging lens is used as a **magnifying glass** it is closer to the object than the focal length. The image is virtual, upright, and magnified. You must look through the lens to see it. The image appears to be further away than the object.

Ray diagram for a magnifying glass

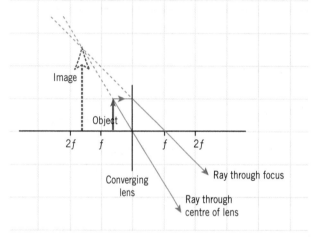

A projector

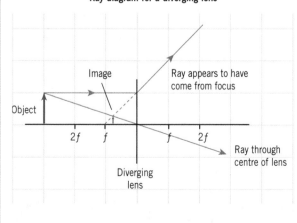

Build Your Understanding

A diverging lens makes a parallel beam diverge so that it appears to have come from the focus. Diverging lenses produce virtual images that are upright and smaller than the object.

Ray diagram for a diverging lens

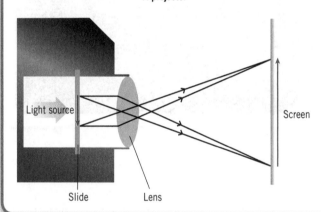

? Test Yourself

1 Which muscles alter the shape of the eye lens?

2 What type of lens is the eye lens?

3 What type of lens is used to correct short sight?

4 Describe the image seen through a magnifying glass.

★ Stretch Yourself

1 What change would laser surgery make to the curvature of the cornea to correct long sight?

2 Describe the image seen by a short sighted person through glasses.

Telescopes and Astronomy

Simple Telescopes

A simple optical telescope has two converging lenses:
- The eyepiece is the lens nearest your eye.
- The objective is the lens nearest the object.

The objective produces a real image. The eyepiece is used as a magnifying glass to magnify this image.

The eyepiece is the more powerful lens – the fatter, or more curved lens.

Build Your Understanding

$$\text{Magnification} = \frac{\text{focal length of objective lens}}{\text{focal length of eyepiece lens}}$$

Example: A telescope has an eyepiece with power 2 D and an objective with power 0.25 D. Focal lengths are $f_e = \frac{1}{2} = 0.5$ m and

$f_o = \frac{1}{0.25} = 4$ m

$\text{Magnification} = \frac{f_o}{f_e} = \frac{4 \text{ m}}{0.5 \text{ m}} = 8$

The telescope has an angular magnification of ×8. Looking through this telescope makes the Moon *appear* eight times (or ×8) bigger or eight times nearer than without the telescope.

Stars are so far away they still look like points through a telescope, but groups of stars are spread out.

Aperture and Brightness

The objective of the telescope gathers all the light rays that enter it and focuses them to a point. This makes the star brighter. You can see stars that are too dim to see with the naked eye. The diameter of the objective is called the **aperture**. To collect radiation from weak or distant sources you need a telescope with a large aperture.

Build Your Understanding

The radiation entering the telescope is diffracted (see page 21) by the aperture. To reduce diffraction the aperture must be much larger than the wavelength of the radiation otherwise the telescope will not give sharp images.

Reflecting Telescopes

Most astronomical telescopes have a **concave mirror**, not a converging lens as the **objective**.

There are lots of different designs, but all have a very large mirror to focus all the light gathered. The eyepiece lens is then positioned to look at the focused light.

The mirror is concave, curved so that parallel rays are focused at a single point.

■ Boost Your Memory

A concave mirror has a caved-in shape.

Light

Build Your Understanding

Advantages of using a mirror:
- A lens focuses different colours (frequencies) at slightly different points, so you do not get such a clear image as with a mirror.
- You can make very large mirrors and support them from behind, but you can't make such large lenses because:
  - The lens would be so heavy it would distort under its own weight.
  - It would be difficult to make the glass even.

Life Elsewhere in the Universe

Astronomers have identified hundreds of distant stars with planets. They have not discovered alien life, existing now or in the past. But because there are so many stars, scientists think it likely that life has evolved somewhere else in the Universe.

The Search for Extra-Terrestrial Intelligence (SETI) project scans radio waves from space for patterns that suggest aliens using radio waves for communication.

Parallax and Parsecs

Parallax was introduced on page 47. Astronomers measure the angle they turn their telescopes in six months to observe a star and record the **angle of parallax**.

The angle of parallax is half of the angle moved against a background of stars in six months. The smaller the angle, the further away the star.

Astronomers measure distance with a unit called the **parsec** (pc).

One parsec is the distance to a star with a parallax angle of one second of arc.
- 1/60th of a degree (1°) is 1 minute of arc (1').
- 1/60th of a minute is 1 second of arc (1").

A **parsec** (3.1×10^{16} m) is similar to a **light year** (9.5×10^{15} m):
- Interstellar distances are typically a few parsecs.
- Intergalactic distances are typically a few megaparsecs (Mpc).

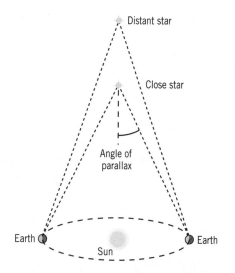

The angle of parallax for a close and distant star

✔ Maximise Your Marks

Get a scientific calculator. They are useful for calculations with large numbers. Make sure you know how to use it and take it to the exam.

❓ Test Yourself

1. Jenna makes a telescope from two lenses. Lens A = 1.5 D and B = 1.0 D. Which is the eyepiece and which is the objective?

2. Which is the best choice of telescope objective, lens diameter 2 cm or 5 cm?

3. Describe the objective of a reflecting telescope.

4. Have scientists detected signs of life on Mars?

⭐ Stretch Yourself

1. Calculate the angular magnification of Jenna's telescope.

2. Two stars have parallax angles 1" and 0.5". Which is closer?

Complete these exam-style questions to test your understanding. Check your answers on page 125. You may wish to answer these questions on a separate piece of paper.

1 This diagram shows a ray of light passing through a prism.

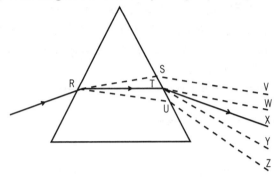

The ray is a ray of red light. Which path would a ray of violet light take through the prism? (1)

A RSV

B RTW

C RTY

D RUZ

2 Light strikes a glass surface at an angle of 40° and is refracted. The refractive index of the glass is 1.5. What is the angle of refraction? (1)

3 The velocity of light in air = 3×10^8 m/s The refractive index of water =1.33. Calculate:

a) The velocity of light in water. (1)

b) The critical angle for a water air boundary. (1)

4 A ray diagram can be used to show how a lens produces an image of an object. Draw a ring round the <u>three</u> things you have to know to be able to draw a ray diagram to find out the **height of the image** and **how far it is from the lens**. (3)

Focal length of the lens.   Thickness of the lens.   Height of the object.

Distance of the object from the lens.   Which way up the image is.

5 A magnifying glass is used to look at a crystal that is 5 mm wide. The magnification is × 4. How wide is the image? (1)

Light

6 A convex lens is used to produce an image of a light bulb on a screen. The lens is positioned to make the image the same size as the light bulb.

a) How would you move the lens to make the image of the light bulb larger? (1)

....

b) Give an example of something that uses a convex lens in this way to make a large image. (1)

....

7 Explain why most astronomical telescopes are reflecting telescopes. (3)

....

....

8 This diagram shows part of the human eye:

a) What is the part labelled A? .... (1)

b) The ciliary muscle contracts. Describe what happens to the eye lens. (1)

....

c) Where are the light rays focused:

i) In a normal eye? .... (1)

ii) In a short sighted eye? .... (1)

d) What type of lens would be used to correct a short sighted eye? (1)

....

....

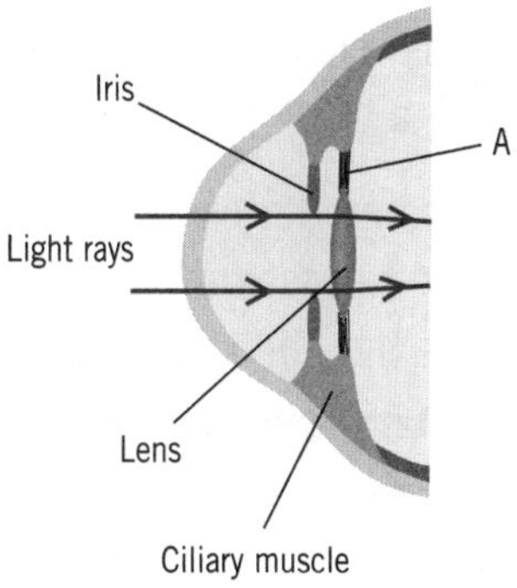

Iris

A

Light rays

Lens

Ciliary muscle

9 A star moves 1 second of arc against the fixed stars in 6 months.

a) What is the angle of parallax? .... (1)

b) Explain why it is 2 parsecs away. .... (1)

10 Describe and explain the similarities and differences between the eye and a digital camera. (In your answer, aim to write between about 8 and 12 lines.) (6)

....

....

....

....

How well did you do?

| 0–7 | Try again | 8–16 | Getting there | 17–21 | Good work | 22–26 | Excellent! |

Electromagnetic Effects

The Motor

There is a force on a current carrying conductor when it is at right angles to a **magnetic field**.

The force is:
- At right angles to both the current and the magnetic field.
- Reversed if the current or the magnetic field is reversed.
- Increased if the current or the magnetic field is increased.
- Zero if the current is parallel to the magnetic field.

The d.c. Motor

When a horizontal loop of wire is placed in a magnetic field the current in one side of the loop produces a force upwards and on the other side it produces a force downwards.

This pair of forces makes the loop rotate to a vertical position. The force is made larger by using a coil with more turns of wire.

The commutator reverses the current direction as the coil passes through the vertical position. This keeps the coil rotating in the same direction.

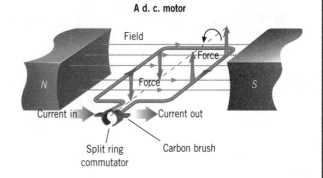

A d. c. motor

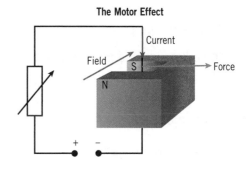

The Motor Effect

Uses of Motors

Motors are widely used. Examples are in:
- Domestic appliances like washing machines and electric drills.
- Electric motor vehicles.
- Hard disk drives.
- DVD and CD players.

Build Your Understanding

The force and motion direction is given by Fleming's left hand rule. The magnets have curved pole pieces to decrease the size of the gap and increase the magnetic field strength.

Some motors have electromagnets instead of permanent magnets, while some have a rotating magnet and fixed coils.

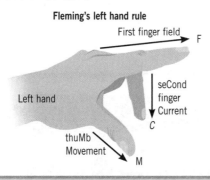

Fleming's left hand rule

Electromagnetic Induction

Electromagnetic induction occurs when a changing magnetic field **induces** a voltage in a conductor. It is used in generators and in transformers. If there is an induced voltage in a coil and the ends are connected to make a complete circuit then a current flows in the coil (see page 7).

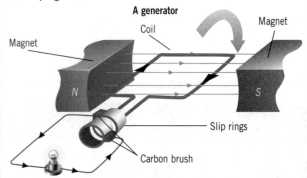

A generator

Induced Voltage

As the coil is rotated voltage is induced in the coil. The output is a.c.

The induced voltage can be increased by:
- Increasing the strength of the magnetic field.
- Increasing the speed of rotation of the magnet or electromagnet.
- Increasing the number of turns on the coil.
- Placing an iron core in the coil.

✓ Maximise Your Marks

If you use the wrong hand rule you will get the wrong direction. The word induced describes the voltage that results from movement and magnetic fields. Always use it for generators, never for motors.

Build Your Understanding

The direction of the current is given by the Fleming's right hand rule.

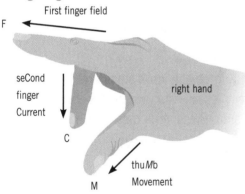

First finger field

F

seCond finger
Current

right hand

C

thuMb
Movement

M

This graph shows how the size of the induced voltage alters as the coil rotates.

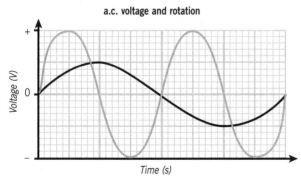

a.c. voltage and rotation

When the coil rotates twice as fast, in other words at twice the frequency, the frequency of the a.c. doubles and the voltage increases. This is shown by the line on the graph.

Transformers

A transformer changes the size of an alternating voltage (see page 12). A transformer is made from a primary coil and a secondary coil wound on an iron core to concentrate the magnetic field in the coils.

An alternating current in the primary coil produces a changing magnetic field. The changing magnetic field induces a voltage in the secondary coil. Step-up transformers have more turns in the secondary coil than the primary coil. Step-down transformers have fewer.

Build Your Understanding

The ratio of the voltages in the coils is:

$$\frac{\text{primary voltage}}{\text{secondary voltage}} = \frac{\text{number of turns in primary coil}}{\text{number of turns in secondary coil}}$$

$$\frac{V_P}{V_S} = \frac{N_P}{N_S}$$

? Test Yourself

1 Which direction does the motor in the diagram rotate?

2 What would happen to the generator if the magnetic field was reversed?

3 What effect does a step-up transformer have on the voltage?

★ Stretch Yourself

A transformer has input voltage 230 V and 920 turns on the primary coil.

1 How many turns are needed on the secondary to give an output voltage of 12 V?

2 If there are 96 turns on the secondary what is the output voltage?

Kinetic Theory

Solids, Liquids and Gases

The **kinetic theory** says that matter is made of particles:

- **Solid** objects are held together by forces between the particles and have a regular shape.
- The particles vibrate more as the object's temperature increases.
- **Liquid** particles have enough energy to break the inter-molecular bonds and slide over each other.
- **Gas** particles have enough energy to separate completely.

The **temperature** increases when the kinetic energy of the particles increases.

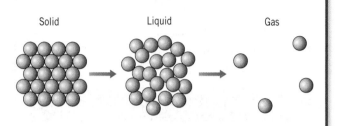

Solid   Liquid   Gas

✓ Maximise Your Marks

A sketch of particles must show that the particle size doesn't change. Solid particles are regularly spaced and touching. The liquid particles are still touching. There are no gaps large enough for another particle to fit in. Gas particles are very widely spaced, so do not draw too many.

Build Your Understanding

Heating an object raises its temperature, except at the melting point and boiling point when heat changes the state, from solid to liquid, or liquid to gas. This is called latent heat.

At the melting point:

- Solids melt when the particles have enough energy to break the bonds holding them together.
- Liquids freeze when the particles lose this energy to the surroundings and form stronger bonds with touching particles.

At the boiling point liquids boil when the particles have enough energy to separate completely. This happens to particles anywhere in the liquid.

Gases condense when the particles lose energy and form weak bonds with other particles.

Liquids evaporate from the surface when the most energetic particles have enough energy to leave the liquid and become a gas. Evaporation happens at any temperature above melting point.

Pressure

When a force acts on a small area it exerts a greater pressure than when it is spread over a large area. Pressure is measured in pascals (Pa).

$$\text{pressure (Pa)} = \frac{\text{force (N)}}{\text{area (m}^2\text{)}}$$

$$1 \text{ Pa} = 1 \text{ N/m}^2$$

A gas is made up of a huge number of atoms or molecules that are constantly moving. The particles randomly collide with other particles and the walls of the container.

The forces exerted by all the particles as they collide with the inside walls of the container add up to give the gas pressure. The pressure is equal in all directions.

Reducing the volume of a gas increases the pressure

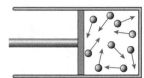

A Model of Pressure

The movement of the particles in the gas is random. They change speed and direction, as they collide with other particles and the walls of the container. This means they change momentum. The force exerted is the rate of change of momentum. The total of all the forces exerted by the particles per square metre of the container walls is the gas pressure.

The Kelvin Temperature Scale

Absolute zero is the temperature at which all particles stop moving. They have no kinetic energy. On the Celsius scale absolute zero is -273ºC.

The **Kelvin temperature scale** is a scale that starts at absolute zero. The unit is the kelvin (K). One kelvin is the same size as one Celsius degree.

To convert between temperature in kelvin and degrees Celsius:
- Temperature in K = temperature in °C +273.
- Temperature in °C = temperature in K –273.

Comparing the kelvin and Celsius temperature scales

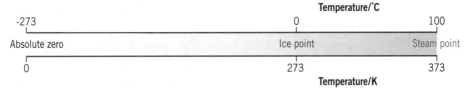

Build Your Understanding

The advantage of using the kelvin temperature scale is that it is an absolute scale starting at zero. This means doubling the temperature doubles the average kinetic energy of the particles. (This is not true if the temperature is measured in °C.)

The temperature is directly proportional to the average kinetic energy of the particles.

❓ Test Yourself

1. What happens to the water molecules in ice as the temperature rises to 3°C?

2. A force of 4000 N acts on an area of 2 m². What is the pressure?

3. What is 100°C in kelvin?

4. What is 195 K in °C?

⭐ Stretch Yourself

1. Gas X has a higher pressure than gas Y. What does this tell you about the particles?

2. The temperature of gas A is 100 K and gas B is 200 K. What does this tell you about the average kinetic energy of their particles?

Further Physics

The Gas Laws

A Fixed Mass of Gas

The gas in a sealed container has mass, measured in kilograms (kg) and volume, measured in metres cubed (m³). The mass is fixed, but by squeezing or heating it you can change the volume.

The pressure is measured in pascals (Pa) and the temperature is measured in kelvin (K).

Pressure and Temperature

When the container has a constant volume and is heated the temperature increases and the pressure increases.

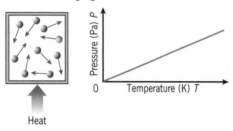

Heating a gas with constant volume

Doubling the temperature doubles the pressure. The graph is a straight line through (0, 0), showing that the pressure is proportional to the temperature.

$$\frac{\text{pressure (Pa)}}{\text{temperature (K)}} = \text{constant}$$

When the gas is heated its temperature rises. The particles have more energy so their average speed is faster:

- A faster particle exerts more force during a collision.
- A faster particle will make more collisions.
- These changes increase the pressure.

Volume and Temperature

When the gas is heated at a constant pressure the temperature increases and the volume also increases to keep the pressure constant.

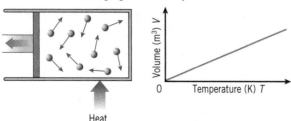

Heating a gas at constant pressure

Doubling the temperature doubles the volume.

The graph is a straight line through (0,0), showing that the volume is proportional to the temperature:

$$\frac{\text{volume (m}^3\text{)}}{\text{temperature (K)}} = \text{constant}$$

The forces exerted by the particles push the sides of the container out to create a larger volume.

Pressure and Volume

When the gas is compressed at a constant temperature the volume decreases and the pressure increases.

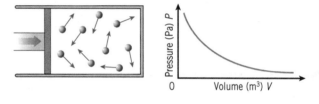

Compressing a gas at constant temperature

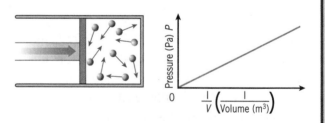

Halving the volume doubles the pressure.

The graph of P against $\frac{1}{V}$ is a straight line through (0,0), showing that the pressure is inversely proportional to the temperature:

pressure × volume (m³) = constant

The particles have the same kinetic energy so if the container is larger the forces exerted by the particles on the sides of the container will be less.

Build Your Understanding

P is pressure, T is temperature and V is volume. At constant volume:

$$\frac{P_1}{T_1} = \frac{P_2}{T_2}$$

Example: Pressure increases from 1×10^5 Pa to 2×10^5 Pa. The temperature was 27°C. Calculate the new temperature T_2.

$T_1 = 273 + 27 = 300$ K

$$T_2 = \frac{P_2 T_1}{P_1} = \frac{2 \times 10^5 \text{ Pa}}{1 \times 10^5 \text{ Pa}} \times 300 \text{ K} = 600 \text{ K or}$$

$600 - 273 = 327$°C

At constant pressure:

$$\frac{V_1}{T_1} = \frac{V_2}{T_2}$$

At constant temperature:

$$P_1 V_1 = P_2 V_2$$

✓ Maximise Your Marks

You must convert temperatures from degrees Celsius to kelvin. Otherwise these equations will not work, because zero pressure and volume occur at -273°C.

⚑ Boost Your Memory

By drawing a table to summarise these relationships, it will help you revise.

An Ideal Gas

These relationships show that if you cool an ideal gas to absolute zero the particles:
- Stop moving completely and have no kinetic energy.
- Make no collisions and exert no pressure.
- Take up no space and have no volume.

A real gas does not behave like this. It condenses to a liquid before reaching absolute zero. These equations work for gases that are above their boiling points, for example nitrogen at room temperature, but not water vapour.

The Gas Law

Collecting together the relationships for pressure temperature and volume.

For a constant mass of gas, with pressure P, temperature T (in kelvin) and volume V:

$$\frac{P_1 V_1}{T_1} = \frac{P_2 V_2}{T_2}$$

Example: A medical oxygen cylinder contains 0.004 m³ of compressed oxygen at a pressure of 1.5×10^7 Pa and has a temperature of 15°C. What volume of oxygen will this give when released from the cylinder?

Atmospheric pressure = 1.01×10^5 Pa and temperature = 20°C.

$V_1 = 0.004$ m³
$V_2 = ?$
$P_1 = 1.5 \times 10^7$ Pa
$P_2 = 1.01 \times 10^5$ Pa

$T_1 = (273 + 15)$ K = 288 K
$T_2 = (273 + 20)$ K = 293 K

$$\frac{1.5 \times 10^7 \text{ Pa} \times 0.004 \text{ m}^3}{288 \text{ K}} = \frac{1.01 \times 10^5 \text{Pa} \times V_2}{293 \text{ K}}$$

$V_2 = 0.604$ m³

❓ Test Yourself

For a fixed mass of ideal gas what happens to:

1 The pressure when you double the volume (temperature is constant).

2 The volume, at one third the pressure (temperature is constant).

3 The temperature in K, when volume is halved (pressure is constant).

4 The particles at absolute zero.

★ Stretch Yourself

1 Tyre pressure at 30°C is 6×10^5 Pa. What is the pressure at 21°C?

2 $P_1 = 1 \times 10^5$ Pa, $V_1 = 0.5$ m³, $T_1 = 290$ K, $P_2 = 1.2 \times 10^5$ Pa, $V_2 = 0.8$ m³. Calculate T_2.

Medical Physics

Ultrasound

Ultrasound is sound with a frequency above the 20 kHz upper threshold of human hearing.

Ultrasound is produced by electronic systems and by animals, for example dogs and bats.

Ultrasound waves are partially reflected at a boundary between two different materials. The rest of the wave is transmitted.

A pulse of ultrasound is transmitted and the returning pulse is detected. The time for a reflection to return is recorded and the distance to the boundary calculated.

Examples are:
- Sonar, to locate the sea floor and shoals of fish.
- Medical diagnosis, for example scans to get information about the eye or the fetus.

Ultrasound scanning of the fetus works like this:
- An ultrasound pulse is emitted from the probe.
- The same probe detects the reflected pulse.
- The two pulses are displayed on an oscilloscope or a computer is used to process all the reflections and produce an image.

- The time for the pulse to travel from the emitter to the fetus and back is recorded.
- The speed of the ultrasound waves is 1500 m/s (speed = distance/time).
- The depth, d, is half the distance the wave travelled:

$$\text{depth} = \frac{\text{speed} \times \text{time}}{2}$$

Other medical uses

Ultrasound can be used to measure the speed of blood flow in the body. When ultrasound waves scan moving objects there is a change in frequency called a Doppler shift.

It can be used for medical treatment, for example to break down kidney stones or gallstones. The vibrations caused by the ultrasound break down the stones.

Advantages over X-ray scans are that ultrasound scans:
- Do not damage living cells or DNA because they are not ionising.
- Can produce images of soft tissue.

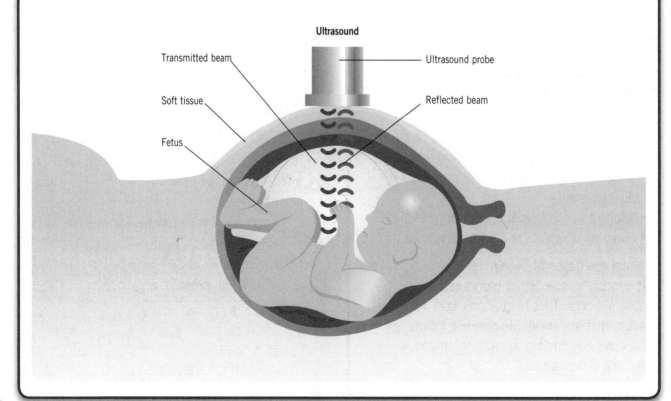

Ultrasound

Transmitted beam · Soft tissue · Fetus · Ultrasound probe · Reflected beam

Build Your Understanding

Example:

Measuring distance with ultrasound

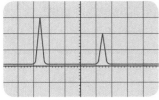

Display screen   Timebase control

The diagram shows the distance between the pulses on the display is 4.0 divisions. The oscilloscope is set to 20 µs per division. So the time for pulse to travel to the boundary and back is 4.0×20 µs $= 80$ µs.

The time for pulse to travel to the boundary $= \frac{1}{2} \times 80$ µs $= 40$ µs.

The average speed of ultrasound waves in the body is 1500 m/s.

distance = speed × time

distance $= 1500$ m/s $\times \dfrac{40\ \mu s}{1\,000\,000}$

$= 0.06$ m $= 6$ cm

✓ Maximise Your Marks

Remember that the distance travelled by the ultrasound pulse is *twice* the depth.

Build β⁺ Decay

The positron is also called a β⁺ particle. It is represented in nuclear equations as $^{0}_{1}e$.

β⁺ decay: a proton in an unstable nucleus changes into a neutron and emits a positron.

Fusion of hydrogen to helium in stars releases positrons: $4\,^{1}_{1}H \rightarrow\, ^{4}_{2}He + 2\,^{0}_{1}e$.

Positron Emission Tomography (PET)

The **positron** is the antiparticle of the electron. It has the same mass and equal, but opposite charge. When it meets an electron they annihilate each other. The mass is converted to energy which is transferred by gamma rays.

To conserve momentum, two equal gamma rays are produced that travel in opposite directions.

An electron and a positron annihilate each other

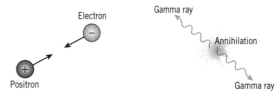

In PET the patient is injected with an isotope that decays by emitting positrons and placed in a scanner that detects gamma rays:

- The isotope decays.
- The positrons travel less than a millimetre before positron-electron annihilation.
- The two gamma rays travel in opposite directions and are detected. Only pairs of gamma rays are used in analysis, any other rays are ignored.
- A picture is built up of the positron concentration.

The isotope used depends on what the scan is for. They are produced by bombarding stable isotopes with protons shortly before being used, as they have short half-lives.

To look at cells that use glucose, like brain cells or tumour cells, a nucleus of fluorine-18 is used to replace an oxygen nucleus in a glucose molecule. After injection the glucose spreads through the body. The body is then scanned.

❓ Test Yourself

1. What is ultrasound?

2. Sonar speed = 1500 m/s, time for pulse to return = 0.08 s. Calculate the water depth.

3. Which particles from protons, positrons and electrons have the same **a)** charge and **b)** mass?

⭐ Stretch Yourself

1. What is the depth of a boundary with pulses five divisions apart on the oscilloscope trace?

2. Write a nuclear equation for the decay of a proton to a neutron.

Practice Questions

 Complete these exam-style questions to test your understanding. Check your answers on page 125. You may wish to answer these questions on a separate piece of paper.

1 Put these steps in order to describe how a transformer works:

   A   The induced voltage makes an alternating current flow in the coil.

   B   The alternating voltage across the primary coil produces a changing magnetic field.

   C   The changing magnetic field induces an alternating voltage in the secondary coil.

   D   The magnetic field changes in the secondary coil.

   E   The magnetic field changes in the iron core.   (4)

2 The diagram shows a coil of wire next to a magnet. A voltmeter is connected to the coil of wire. The magnet rotates.

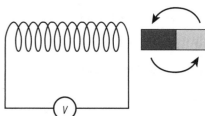

As the magnet completes one full turn:

   a)   Describe what happens to the reading on the voltmeter. (1)

   b)   What is the name of this effect?   (1)

   c)   What will happen if the magnet is rotated more quickly?   (1)

3 A transformer is made of two coils of wire on an iron core. An alternating input voltage causes an alternating output voltage.

   a)   A 3 V battery is connected to the input coil. Explain why there is no voltage across the output coil. (1)

   b)   A 3 V a.c. source is connected to the input coil. Explain why there is a voltage across the output coil. (1)

   c)   The input coil has 200 turns. When the input voltage is 3 V the output voltage is 12 V. Calculate the number of turns in the output coil.   (1)

   d)   Explain the difference between a step-up and a step-down transformer.   (1)

Further Physics

4 The diagram shows a d.c. motor:

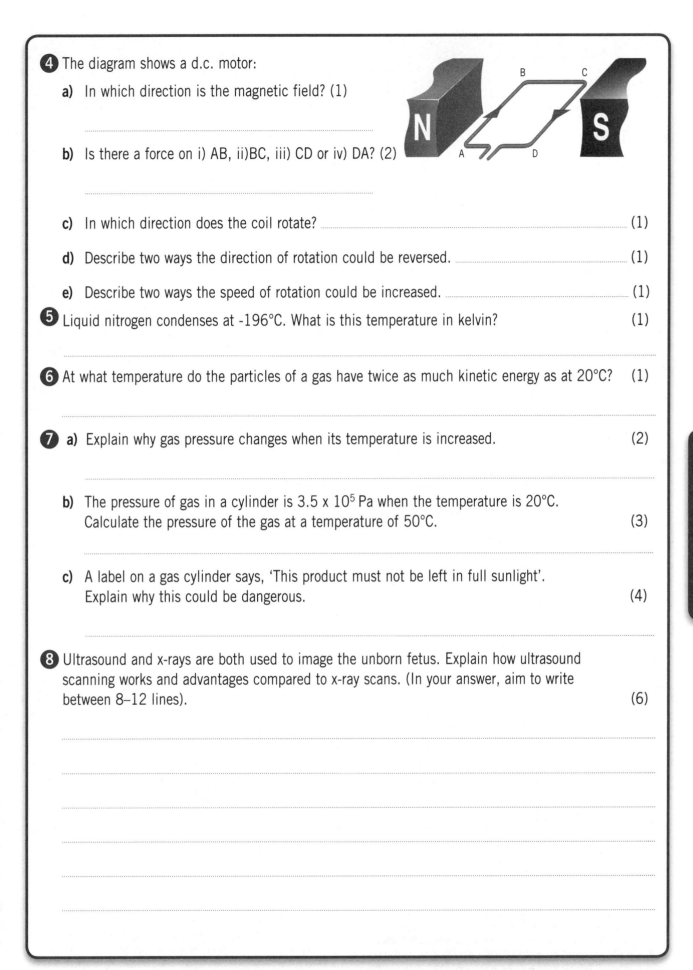

a) In which direction is the magnetic field? (1)

b) Is there a force on i) AB, ii) BC, iii) CD or iv) DA? (2)

c) In which direction does the coil rotate? .... (1)

d) Describe two ways the direction of rotation could be reversed. .... (1)

e) Describe two ways the speed of rotation could be increased. .... (1)

5 Liquid nitrogen condenses at -196°C. What is this temperature in kelvin? (1)

6 At what temperature do the particles of a gas have twice as much kinetic energy as at 20°C? (1)

7 a) Explain why gas pressure changes when its temperature is increased. (2)

b) The pressure of gas in a cylinder is 3.5×10^5 Pa when the temperature is 20°C. Calculate the pressure of the gas at a temperature of 50°C. (3)

c) A label on a gas cylinder says, 'This product must not be left in full sunlight'. Explain why this could be dangerous. (4)

8 Ultrasound and x-rays are both used to image the unborn fetus. Explain how ultrasound scanning works and advantages compared to x-ray scans. (In your answer, aim to write between 8–12 lines). (6)

How well did you do?

| 0–6 | Try again | 7–15 | Getting there | 16–25 | Good work | 26–33 | Excellent! |

Answers

Topic 1 Energy

The Electricity Supply
Test Yourself
1. Electricity has to be made from another, primary, energy source.
2. **a)** Arrow 100% = 1300 MJ on left splits into useful energy = 755 MJ and wasted energy = 545 MJ on the right; **b)** (755 ÷ 1300) × 100 = 58%
3. (8 ÷ 60) × 100 = 13%

Stretch Yourself
1. **a)** 3500 kJ = 100% on left splits into useful = 1190 kJ = 34% two wasted arrows heat in power station = 2010 kJ = 57% overhead cables = 300 kJ=9%; **b)** (1190 ÷ 3500) × 100 = 34%
2. When the primary energy resource is used to generate electricity energy is wasted as heat in the power station and in cables in the National Grid. If the primary resource is burned in your home all the heat is useful.

Generating Electricity
Test Yourself
1. Coal is burned, but the uranium/plutonium nucleus splits in a nuclear reaction (fission).
2. Uranium or plutonium.
3. Chicken manure and willow.

Stretch Yourself
1. Examples are: *For nuclear:* No carbon dioxide emissions emitted by nuclear fission; very large amounts of energy from each power station and small amount of fuel; *Against nuclear:* Radioactive waste produced which must be safely stored; target for terrorist activity; natural disasters can cause leaks of radioactive material.
   For fossil fuel: Large amount of energy from a few power stations; no highly radioactive waste. *Against fossil fuel:* Carbon dioxide emitted causes global warming and may cause climate change.
   Other: Pollution causes acid rain; limited supply will eventually run out; mining and drilling operations can cause environmental damage.
2. *Advantages:* Carbon dioxide released is matched by carbon dioxide taken up during life of fuel, so there is no net change; a way of using up waste products like manure, or methane. *Disadvantages:* Large amount of land required to grow some crops – maybe conflict with food crops or wildlife; supply needs to be planned – e.g. willow planted.

Renewable Sources of Energy
Test Yourself
1. Rain or snow fills lakes or reservoirs high up in the mountains. This water falls through fast rivers or pipes gaining kinetic energy. It is used to turn the turbines at the hydroelectric power station, which turn the generators to generate electricity.
2. Examples of suitable choices: *Wind turbines:* Offshore windfarms can make use of the high winds around the west coast. *Wave generators:* Once suitable technology has been developed we have lots of sites where ocean waves can be exploited by wave generators. *Solar cells:* Newer technology means that solar cells can produce useful amounts of energy in the UK.
3. Examples of unsuitable choices: *Hydroelectric power* because there are no high mountains, so we only have very small amounts of HEP. *Tidal energy* because there are no large tidal changes. The Severn estuary would be the best site, but is an important site for wildlife.

Stretch Yourself
1. **a)** No carbon dioxide emitted; renewable-fuel will not run out; some good sites round the coast. **b)** Reasons could include: Winds and waves very variable; often little wind when there is very cold weather; lots of space required to generate a small amount of electricity; lots of overhead power lines needed to connect them all to the grid; some people consider wind turbines an eyesore and they can be noisy; it is best to have a mix of energy generation methods to avoid a complete loss of production, e.g. if there is a very damaging storm.
2. **a)** A solar panel would not need connecting to the grid and can charge batteries during the day for use at night. A light does not require much energy. **b)** This requires a lot of electricity. An HEP station could be built because the area is mountainous, and it would generate a large amount.

Electrical Energy and Power
Test Yourself
1. Watts (W)
2. 11.5 W
3. 2 kW × 0.1 h = 0.2 kWh
4. 800 W × 60 s = 48000 J

Stretch Yourself
1. 1000 W ÷ 230 V = 4.3 A
2. Units = 1 kW × 0.25 h = 0.25 kWh; cost = 0.25 × 12p = 3p

Electricity Matters
Test Yourself
1. a.c. is alternating current (it reverses direction); d.c. is direct current (it has a steady current).
2. The National Grid is a distribution network of cables and overhead power lines and transformers that links power stations with users of electricity.
3. *Any two similarities*, e.g. both have generators, turbines, boilers, both heat water to steam. *Any two differences*, e.g. nuclear: small amount of nuclear fuel (uranium/plutonium) coal: large amount of coal; nuclear: no CO_2, but produces radioactive waste; coal: CO_2 lot of waste, but not radioactive; nuclear: nuclear reaction/fission releases a lot of energy; coal: burning/combustion of coal, less energy released.
4. Radioactive waste is dangerous because it emits ionising radiation which can kill or damage living cells or cause them to turn cancerous.

Stretch Yourself
1. A step-down transformer.
2. Transmitting at a high voltage means that a much smaller current can be used, for the same amount of power. This means that there is less heat lost in the cables, which saves a lot of energy, as there are hundreds of miles of cables.

Particles and Heat Transfer
Test Yourself
1. E = 1 kg × 4200 J/kg°C × 2°C = 8400 J
2. Diagram similar to convection current in house shown on page 14. Hot water rises, cold water sinks.
3. £2000 ÷ £100 per year = 20 years.

Stretch Yourself
1. 4400 J = 0.5 kg × c × 10°C; c = 4400 J ÷ (0.5 kg × 10°C) = 880 J/kg°C
2. It traps the air in an insulating foam, so that conduction and convection are stopped.

Practice Questions
1 C; 2 D
3. **a)** Uranium (and/or plutonium); **b)** Split the nucleus of an atom into two or more parts; **c)** Heats water to steam which turns turbines; **d)** Water turns turbines directly; e) Advantage: e.g. No waste, renewable energy; Disadvantage: depends on steady water supply, or may require a dam to flood a valley.
4. **a)** (755 MJ ÷ 1300 MJ) ×100 = 58%; **b)** Burns gas not coal, uses hot gas and steam, coal uses steam only, or more efficient (58 per cent compared to about 40 per cent); **c)** Produces carbon dioxide which causes global warming or produces pollution, e.g. sulfur dioxide, or is a fossil fuel so will run out.
5. **a)** 3 × 2 = 6 units; **b)** The light; **c)** 6 units × 10p = 60p
6. *Advantages* of a coal-fired power station are that it can produce a lot more electricity, hundreds, or a few thousand megawatts, but even the largest wind turbines only produce a few megawatts. To replace one power station would need several windfarms, which would require a large amount of land. The windiest places are often areas of natural beauty and lot of people think windfarms are an eyesore. Some people say they are noisy, and harmful to birds and bats. Power stations can be built in less attractive places. Power stations are also more reliable, as when the wind drops, wind turbines generate no power. The *disadvantages* of coal-fired power stations are that coal is a fossil fuel and will eventually run out, whereas wind is renewable. Burning coal produces carbon dioxide which is a greenhouse gas and contributes to global warming, and there is other

Answers

waste, like smoke and ash. Wind turbines produce no waste gases, smoke or ash. Coal has to be mined, bought and transported to the power station. The wind is free.

Topic 2 Waves

Describing Waves
Test Yourself
1 Diagram – as shown in diagram on page 18. Amplitude – crest to undisturbed position (or trough to undisturbed position); wavelength – any point on a wave at exactly the same point.
2 It has oscillations parallel to the direction of travel, not at right angles.
3 256 Hz × 1.3 m = 333 m/s

Stretch Yourself
1 1 ÷ 4 = 0.25 Hz
2 $2f = v \lambda \div f = 3 \times 10^8$ m/s ÷ 3 m = 1×10^8 Hz ÷ 100 000 000 Hz

Wave Behaviour
Test Yourself
1 30°
2 The angle of refraction is smaller.
3 As diagram on page 21 **a)** top boundary, **b)** bottom boundary (air is less dense than water.)
4

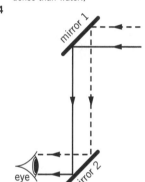

Stretch Yourself
1 **a)** Like diagram c) on page 21; **b)** Like diagram a) on page 21; **c)** Like diagram b) on page 21.

Seismic Waves and the Earth
Test Yourself
1 P-waves.
2 A seismometer is an instrument for detecting seismic waves from, e.g. earthquakes.
3 It's made up of a number of plates floating on the mantle.
4 The time delay between P-waves and S-waves arriving depends on how far away the earthquake is.

Stretch Yourself
1 **a)** P-waves are longitudinal; **b)** S-waves are transverse.
2 Because there is an S-wave shadow zone, S-waves cannot travel through liquids.

Practice Questions
1 A; **2** D
3 A wave is the movement of a **disturbance** through a **medium**. Waves transfer **energy** but not **matter**. A wave is caused by something that **vibrates**.
4 P-waves are longitudinal, S-waves are transverse. P-waves travel through liquids, S-waves do not (or P-waves travel faster than S-waves).
5 False; true; false.
6 Refraction is the slowing down or speeding up of waves when they enter a different medium. The water waves slow down when they enter shallow water and if they are travelling at an angle to the boundary they will change direction.
7 a) 0.1 Hz b) v = fλ c) 20 m/s
8 **a)** P-waves, they arrive first; **b)** The P-waves and the S-waves would arrive later, because they have further to travel. The time between the P and S waves would be longer, because the S-waves will fall further behind the P-waves the further they travel; **c)** There would be no S-waves because the outer core is liquid and they cannot travel through liquid so only P-waves will reach the far side of the Earth.

9 *Similarities* are that they are both waves which spread out from a source and transmit energy but not matter. They can both be used to carry information. They can be reflected at refracted boundaries between different materials. They are diffracted by gaps and round corners. They both show interference effects. The *differences* are that sound is a longitudinal wave and travels as compression and rarefactions passing through a medium, it cannot travel in a vacuum, whereas light is a transverse wave and is oscillations of an electric and magnetic field so it can travel through a vacuum. Sound travels at a speed of about 300 m/s in air whereas light is a million times faster at 300 000 000 m/s in a vacuum (slightly less in air). Sound is diffracted through doorways as the wavelength is about a metre, whereas light has a wavelength of about 5 x 10^{-7} m and is only diffracted through tiny gaps of about that size.

Topic 3 Electromagnetic Waves

The Electromagnetic Spectrum
Test Yourself
1 Two from: transverse, travel at 3×10^8 m/s in a vacuum, can travel through a vacuum, are oscillations of a magnetic and electric field.
2 Infrared have lower frequency or longer wavelength, or lower energy than ultraviolet. Or ultraviolet are ionising, infrared are not.
3 Infrared.
4 Ultraviolet, X-rays, gamma rays.

Stretch Yourself
1 Source B has the most energy (higher frequency so higher energy).
2 A = 3×10^8 ÷ 3×10^9 m = 0.1 m or 10 cm; B = 3×10^8 ÷ 30×10^9 m = 0.01 m or 1 cm

Light, Radio Waves and Microwaves
Test Yourself
1 The image is smaller and upside down compared to the object.
2 Glass or plastic because they do not absorb microwaves.
3 850 W × 10 s = 8500 J
4 They are absorbed by water molecules and heat them up. The human body contains a lot of water, or human cells contain water.

Stretch Yourself
1 If they are sent through the atmosphere between a transmitter and a receiver, the amount received will depend on whether it is raining, this could be used to track rainfall.

Wireless Communications 1
Test Yourself
1 **a)** Radio waves; **b)** Microwaves.
2 Need to be in 'line of sight' of the next one.
3 An analogue signal is continuously changing but digital has just two values – a labelled diagram to show this is acceptable.
4 Unwanted signals that are picked up and added to the signal.

Stretch Yourself
1 So that diffraction does not make the beam spread out.
2 **a)** Noise; **b)** Because digital signals can be cleaned up; the 0 and 1 values can be restored; the noise; signals that produce the white specks can be removed.

Wireless Communications 2
Test Yourself
1 Examples could include: We can receive phone calls and e-mail 24-hours a day; no wires are needed to connect laptops to the internet, or for mobile phones or radio; communication with wireless technology is portable and convenient.
2 *Benefit*: staying in contact with family/friends or emergency use. *Risk*: possible unknown long-term health risk, or risk of distraction when crossing road/driving.
3 Graph showing a positive correlation – as hay fever increases sales of ice cream increase, or as sales of ice cream increase, hay fever increases.
4 A small heating effect (or vibrate molecules).

Stretch Yourself
1 Because they will be using them a lot over their lifetime and it is too early to be sure there are no long-term effects.
2 **a)** Factor = distance travelled in glass. Outcome = microwaves absorbed (fall in intensity detected); **b)** The greater the distance the more microwaves are absorbed (or lower intensity detected); **c)** They would all affect the intensity of microwaves received by the detector, so it would not be possible to tell what effect changing the distance in glass would have.

Infrared

Test Yourself

1 Our skin feels infrared as heat.
2 The infrared radiation from the wire is absorbed by the molecules on the surface of the bread which gets hot. Conduction then carries this heat through the rest of the bread.
3 It is cooler so the infrared radiation has a lower frequency.
4 It is a digital code of 0s and 1s so the pattern of 0s and 1s would be different.

Stretch Yourself

1 It is a thin beam (not divergent); it is one frequency, e.g. one colour red – the light bulb is lots of frequencies; the waves are all in phase (in step) they are not from the light bulb
2 Less interference; signals can be multiplexed and many sent at once.

The Ionising Radiations

Test Yourself

1 It removes electrons from atoms making them more likely to take part in chemical reactions.
2 Contaminated is when you swallow/breathe-in/get covered by radioactive material. Irradiation is when the rays from radioactive material reach your body.
3 Because the ionising radiation is damaging and the benefit does not outweigh the risk.
4 Ultraviolet

Stretch Yourself

1 Walls absorb X-rays, so they will not be exposed.
2 Ultraviolet causes skin to make vitamin D, but also causes skin cancer.

The Atmosphere

Test Yourself

1 Carbon dioxide, water vapour and methane.
2 Because of burning fossil fuels/clearing forests.
3 Ozone

Stretch Yourself

1 Whether human activities are responsible.
2 CFCs. They have been banned by governments all over the world.

Practice Questions

1 Microwaves – communication with satellites
  Radio waves – TV broadcasts
  Infrared – optical fibre communications
  X-rays – medical scans
2 500 nm – visible light
  0.01 mm – infrared
  3 cm – microwaves
  300 m – radio waves
3 Intensity = energy per square metre per second, so the intensity = 100 W/m^2
4 They can receive microwaves from the satellite which is above them, but not radio waves or microwaves from transmitters as they are blocked by the hills.
5 An analogue signal is continuously changing. A digital signal has only two values, 0 and 1 (or on and off), and is a series of pulses.
6 a) Same numbers at the start and end for each group (phone users and non-users). Very small sample – results could be just random chance; b) Bigger sample will give more meaningful results – but by the end of the trial sample not using phone is small; c) If the 200 who start to use a phone are removed there is a sample of 800, 400 who use a phone and 400 who don't; d) Not really – it could take longer for effects to be seen; e) Texting does not expose the brain to so many microwaves, so could make using a mobile phone look safer than it is. A good study would need to know more about the actual phone use.
7 TV signals can be added to a radio wave (called a carrier wave) and broadcast from a transmitter mast. The radio waves travel through the atmosphere and are diffracted through valleys and around hills. They can be reflected from the ionosphere. They are received by a TV aerial and produce electric currents in the aerial. The signal is removed from the carrier radio wave by the TV receiver. Another way is to add the signal to a microwave carrier wave and send it to a satellite which receives the signal and retransmits it back to Earth. The signal is received by satellite receiver dishes and the microwave carrier wave is removed by the satellite receiver.

Topic 4 Beyond the Earth

The Solar System

Test Yourself

1 Jupiter
2 An asteroid is a rock that orbits the Sun between Mars and Jupiter. A comet is a rock and ice that orbits the Sun in a very elliptical orbit, spending most of the time further away from the Sun than Pluto.
3 The Sun is at the centre of the Solar System. The planets orbit the Sun and moons orbit the planets.
4 The Sun

Stretch Yourself

1 Gravity
2 The orbit of a comet is very elliptical going close to the Sun and further away than Pluto. The orbit of the Earth is almost circular and it stays roughly the same distance from the Sun.

Space Exploration

Test Yourself

1 Telescopes and unmanned space probes.
2 a) Atmospheric effects (pollution, clouds, dust).
  b) Expense of launch, or difficult to maintain or repair.
3 The Earth collided with a planet – debris formed the Moon.

Stretch Yourself

1 Examples: manned are more interesting/inspirational for people. People can respond to changes or make repairs and unmanned space probes can't. But manned missions cost more and lives are at risk if something goes wrong.
2 The Moon has no iron core and its rocks are the same as Earth rocks. It has no volcanic activity, but has igneous rocks.

A Sense of Scale

Test Yourself

1 Comet, Moon, Earth, Sun, Milky Way, Universe.
2 About 10 years.
3 Lamp post.

Stretch Yourself

1 You don't know how bright the star really is.
2 Distant stars don't appear to move when the Earth changes its position in six months.

Stars

Test Yourself

1 The outer layers of a red giant star that has used up its helium and are drifting away into space in all directions.
2 A brown or black dwarf.
3 A supernova.
4 A red supergiant is more massive, and after helium fusion is complete it goes on to fuse other nuclei until it produces iron. Finally, it explodes as a supernova. A red giant stops after helium fusion is finished producing.

Stretch Yourself

1 By nuclear fusion, inside red giant and red supergiant stars.
2 Because no light can escape it as there is such a strong gravitational attraction.

Galaxies and Red-Shift

Test Yourself

1 They are all moving away from us and each other. Those furthest away are moving fastest.
2 When scientists evaluate each others' work.
3 Yes
4 Towards the red end of the spectrum.

Stretch Yourself

1 That the Andromeda galaxy is moving towards us.
2 Elements present in a star, and whether the star is moving towards or away from us (and if so, how fast).

Expanding Universe and Big Bang

Test Yourself

1 The Universe is expanding.
2 The Big Bang was 14 thousand million years ago.
3 The cosmic microwave background radiation (CMBR).
4 What happened before the Big Bang?

Stretch Yourself

1 CMBR left over from the Big Bang.
2 The discovery of the CMBR because other theories could not explain it.

Practice Questions

1 Between Mars and Jupiter. The asteroids have not formed a planet because of the gravity of Jupiter.
2 A = Sun, B = Earth, C = Saturn, D = Mars, E = Uranus, F = Venus, G = Neptune, H = Pluto, I = Jupiter, J = Mercury. Place an X somewhere between Mars and Jupiter
3 **a)** = F; **b)** = F; **c)** = T; **d)** = T; **e)** = T; **f)** = T
4 Pluto, Jupiter, Sun, Earth's orbit, Solar System, Milky Way.
5 **a)** Alpha would be dimmer than Beta; **b)** Beta would move more.
6 C
7 C
8 The star is getting further away from us.
9 Helium and hydrogen.
10 The Big Bang theory is that the Universe began as a small point about 14 thousand million years ago. All space and matter expanded out from the point, called the Big Bang, and it is still expanding today. The evidence comes firstly, from the red-shift of electromagnetic radiation from all the distant galaxies and the fact that the red-shift is larger the further away the galaxy is from us. This shows that all the galaxies are getting further away and those furthest away are moving away fastest, which shows that space is expanding. Scientists predicted that radiation from the big bang should still be present in all directions and that this would be found in the microwave region of the spectrum. In the 1960s scientists discovered the cosmic microwave background radiation and this evidence lead to the Big Bang being accepted by most scientists.

Topic 5 Forces and Motion

Distance, Speed and Velocity
Test Yourself

1 Instantaneous speed is the speed at any instant. Average speed is calculated from the total distance travelled divided by time taken and the speed may have changed several times during the journey time.
2 $(100\text{ m} - 0\text{ m}) \div (30\text{ s} - 0\text{ s}) = 3.33$ m/s
3 Speed = 288 km $\div$ 3 h = 96 km/h
4 $t = d \div s$: $t = 1.44$ km $\div$ 12 m/s = 1440 m $\div$ 12 m/s = 120 s (= 2 minutes)

Stretch Yourself

1 **a)** 0.9 km; **b)** Between 1.5 and 2 hours (or on the way home from the shop); **c)** Because the graph is steepest at this point.

Speed, Velocity and Acceleration
Test Yourself

1 24 m/s $\div$ 8 s = 3 m/s^2
2 The object is travelling at a steady/constant speed.
3 **a)** 0 m/s^2
   b) $(0-20)$ m/s $\div$ $(80-60)$ s = -1 m/s^2

Stretch Yourself

1 Line goes below x axis similar to last part of the graph for the rolling ball, then horizontal, then straight line sloping upwards to the x axis again.
2 **a)** 20 m/s $\times$ (60 s – 40 s) = 400 m; **b)** ½ $\times$ 20 m/s $\times$ (80 s – 60 s) = 200m

Forces
Test Yourself

1 2 kg $\times$ 10 N/kg = 20 N
2 3000 N – 900 N = 2100 N
3 0 N because there is no change in motion.

Stretch Yourself

1 **a)** 0.1 kg $\times$ 10 kg/N = 1 N; **b)** 1/6 $\times$ 1 N = 0.17 N
2 **a)** 2500 N – 800 N – 1700 N = 0 N; **b)** It will continue travelling at the same speed, because forces balanced/resultant = 0.

Acceleration and Momentum
Test Yourself

1 1200 kg $\times$ 3 m/s^2 = 3600 N
2 2 kg $\times$ 5 m/s = 10 kgm/s (or 10 Ns)
3 0.2 kg $\times$ $(-8$ m/s$)$ = -1.6 kgm/s (or -1.6 Ns)
4 50 N $\times$12 s = 600 Ns

Stretch Yourself

1 **a)** Force = (1000 kg $\times$ 12 m/s) $\div$ 0.002 s = 6 000 000 N; **b)** Stopping force = (1000 kg $\times$ 12 m/s) $\div$ 0.5 s = 24 000 N; **c)** There would be less damage to the truck.

Pairs of Forces: Action and Reaction
Test Yourself

1 The wall pushes on the girl with a force of 5 N.
2 Because there is little friction to push back on your foot and send you forward.
3 It pushes exhaust gases out the back, so there is an equal and opposite force pushing the rocket forward.
4 The balloon pushes air out of the back and an equal and opposite force on the balloon pushes it forward.

Stretch Yourself

1 1st pair: The weight of the book and the equal and opposite gravitational attraction of the book on the Earth. 2nd pair: The contact force of the book pushing down on the table and the equal and opposite force of the table pushing up on the book.
2 They both stop (momentum = 0), so before the collision the momentum of the car is equal and opposite to the momentum of the truck.
   Car: mv = 0.5 kg $\times$ 4 m/s = 2 kgm/s
   Truck: MV = 2 kg $\times$ V V = -2 kgm/s $\div$ 2 kg = -1 m/s (speed = 1 m/s)

Work and Energy
Test Yourself

1 1000 N $\times$ 200 m = 200 000 J
2 12000 N $\times$ 30 m = 360 000 J
3 ½ $\times$ 900 kg $\times$ (15 m/s)^2 = 101 250 J
4 If there is friction some of the energy is not transferred from GPE to KE.

Stretch Yourself

1 **a)** 8000 N x 50 m = 400 000 J; **b)** 400 000 J

Energy and Power
Test Yourself

1 When the frictional forces are so small they can be ignored, or in a vacuum.
2 The GPE is transferred to heating the air and the object.
3 Any of listed examples, e.g. driver has been drinking or poor condition of tyres.

Stretch Yourself

1 The cyclist is doing work against friction forces the energy is transferred as heat to the bicycle, road and air.
2 3 $\times$ 30 mph so thinking distance = 3 $\times$ 9 m braking distance = 9 $\times$ 14 m stopping distance = 27 + 126 = 153 m

Practice Questions

1 **a)** i) D; ii) C; **b)** 20 m/s $\times$ 20 s = 400 m; **c)** 50 m/s
2 **a)** It accelerates at a steady rate from 0 m/s to 14.0 m/s in 10 seconds. It travels at a steady speed for 25 seconds. It slows down at a steady rate to 8.5m/s over 15 seconds; **b) i)** $(14 - 0)$ m/s $\div$ 10 s = 1.4 m/s^2; **ii)** 14 m/s; **iii)** (8.5 m/s – 12 m/s) $\div$ 10 s = -3.7 m/s $\div$ 10 s = -0.37 m/s^2 [or (8.3 m/s–14 m/s) $\div$ 15 s = -0.38 m/s^2] (answers between 0.37 m/s^2 and 0.4 m/s^2 are acceptable); **c)** The distance travelled during the first 10 seconds; **d) i)** ½ (14 m/s $\times$ 10 s) = 70 m; **ii)** 72.5 m + (14 m/s $\times$ 10 s) = 75 m + 290 m = 365 m
3 **a)** A = B = 1040; C = D = 2480; E = 15; F = 116; G = 86; H = 620
   b) Higher speed is causing more serious injuries. The reason (mechanism) is that when the speed is higher the stopping force is greater. This is shown by the stopping force increasing from 15 kN to 86 kN when wearing a seatbelt and when speed increases from 30 mph to 70 mph, and from 116 kN to 620 kN when not wearing a seatbelt; **c)** The injuries are reduced because the force is less, and this is because the seatbelt increases the time it takes to stop, e.g. at 30 mph wearing a seatbelt increases the time from 0.009 s to 0.069 s and this reduces the force from 116 kN to 15 kN (or at 70 mph it increase the time from 0.004 s to 0.029 s and this reduces the force from 620 kN to 86 kN); **d)** *Benefits*: less chances of injury and death in an accident, less chance of injuring or killing other occupants of car. *Risks*: may take extra time to escape vehicle, may drive faster (feel safer) and more chance of injuring other road users.
4 A2; B1; C4; D3
5 In a collision the stopping force on the driver causes the damage, as he (or she) suddenly loses all his (or her) momentum. The force is equal to the

rate of change of momentum. To reduce the force, the rate of change of momentum is reduced by increasing the time over which the change takes place. The driver is brought to a stop more slowly. Crumple zones, seat belts and air bags all increase the time for the driver to stop, so they reduce the stopping force, reducing injury. Metal crumple zones at the front and the back of the car crumple, so that the time from the start of the collision to the final stop is longer. Seat-belts stretch slightly, so that the driver moves forward while slowing down. The time to stop is much longer than if the driver hits the windscreen. An airbag inflates in a collision and then compresses as the driver hits it, so that he (or she) moves further and takes longer to stop than if he (or she) hits the steering wheel.

Topic 6 Electricity

Electrostatic Effects
Test Yourself
1 They have opposite charges, which attract.
2 Electrons
3 Two – positive and negative.
4 Nylon is a better insulator. Some of the charges are conducted away through the wool carpet.

Stretch Yourself
1 **a)** They are opposite; **b)** Repel because they are the same, and they would get the same charge.
2 So that there is not a build up of charge which could cause a spark. A spark could ignite the fuel.

Uses of Electrostatics
Test Yourself
1 The grids will attract (or repel) electrons from the smoke particles, which will be left with the same charge as the grid, so the plates need to have the opposite charge to attract the charged smoke particles.
2 They would be repelled and would not stick to the car body.
3 Better coverage or less wastage.
4 The shock could stop their heart.

Stretch Yourself
1 So that they will charge the smoke particles that come close to them, by attracting or repelling electrons.
2 Electrons flow from earth to neutralise the positive charge of the paint drops landing on the metal.

Electric Circuits
Test Yourself
1 **a)** 0.8 A; **b)** 0.8 A
2 **a)** 400 mA; **b)** 900 mA

Stretch Yourself
1 20 C
2 reading A = reading B + reading C

Voltage or Potential Difference
Test Yourself
1 **a)** 2 V; **b)** No, current is the same everywhere in a series circuit.
2 **a)** 9 V; **b)** Only if the lamps are identical. They might draw a different current and have different brightness.

Stretch Yourself
1 $9 \text{ V} \times 10 \text{ C} = 90 \text{ J}$
2 **a)** $5 \times 1.5 \text{ V} = 7.5 \text{ V}$; **b)** 1.5 V
3 They will last longer – there is five times as much stored charge..

Resistance and Resistors
Test Yourself
1 90 Ω
2 0.06 A
3 **a)** 600 Ω; **b)** $R_3 = 300$ Ω

Stretch Yourself
1 The free electrons collide with the stationary positive ions giving them energy so they vibrate more, which means they are hotter.
2 Yes, because $V/I = 150$ Ω in each case, so R is constant when V and I change, V is proportional to I.

Special Resistors
Test Yourself
1 Its temperature increases and so resistance increases.
2 It has too high a resistance and would get hot when current flowed.
3 Its resistance decreases when temperature increases.
4 A diode – it only allows current to pass in one direction.

Stretch Yourself
1 **a)** $V_1 = 5 \text{ V} \times \dfrac{300 \text{ }\Omega}{(300 \text{ }\Omega + 200 \text{ }\Omega)} = 3 \text{ V}$
  b) $V_2 = 5 \text{ V} \times \dfrac{200 \text{ }\Omega}{(300 \text{ }\Omega + 200 \text{ }\Omega)} = 2 \text{ V or } 5 \text{ V} - 3 \text{ V} = 2 \text{ V}$
2 **a)** decreases; **b)** decreases; **c)** increases.

The Mains Supply
Test Yourself
1 It won't melt if the current gets too high – other wires may melt first causing a fire.
2 If the appliance becomes live it won't melt the fuse, so someone touching it could get a fatal shock
3 **a)** $2500 \text{ W} \div 230 \text{ V} = 10.9 \text{ A}$; **b)** $9 \text{ W} \div 230 \text{ V} = 0.04 \text{ A}$; **c)** $300 \text{ W} \div 230 \text{ V} = 1.3 \text{ A}$
4 **a)** 13 A; **b)** 3 A; **c)** 3 A

Stretch Yourself
1 If they get wet/wires get cut/other fault the power supply will be cut off. When the fault is fixed the power can be switched back on without replacing the fuse.
2 **a)** $(0.5 \text{ A})^2 \times 100 \text{ }\Omega = 25 \text{ W}$; **b)** $(1 \text{ A})^2 \times 100 \text{ }\Omega = 100 \text{ W}$

Practice Questions
1 **a)** They will repel because they have the same charge and like charges repel; **b) i)** Rod A has the opposite charge to the Perspex. **ii)** Rod B has the same charge as the Perspex.
2 **a)** B; **b)** C
3 **a)** An ammeter; **b)** A voltmeter.
c)

	Current (mA)	Voltage (V)
Lamp	50	6
Resistor	50	3

4 **a)** 20 mA; **b)** $20 \text{ mA} \times 60 \text{ }\Omega = 1.2 \text{ V}$; **c)** $3 \text{ V} - 1.2 \text{ V} = 1.8 \text{ V}$; **d)** $50 \text{ mA} - 20 \text{ mA} = 30 \text{ mA}$; **e)** $3 \text{ V} \div 30 \text{ mA} = 100 \text{ }\Omega$
5 **a)** = 500 Ω; **b)** = 5 V; **c)** = 10 Ω; **d)** = 0.15 V; **e)** = dark; **f)** = 1000 Ω; **g)** = resistor; **h)** = variable; **i)** = resistor
6 $500 \text{ W} \div 230 \text{ V} = 2.17 \text{ A}$ so 3 A (B)
7 **a)** $8000 \text{ W} \div 230 \text{ V} = 34.8 \text{ A}$; **b)** A very high current so this would heat cables, thick cables have lower resistance so less heating.
8 The paint droplets are given a positive electric charge by the spray gun. The metal car part is given a negative electric charge by connecting it to an electric voltage. (Or, the metal is earthed, so negative charge is attracted from earth towards the positive paint drops, making the car part negatively charged.) The positively charged paint droplets will be attracted to the negatively charged metal car part because unlike charges attract. The attractive force will pull the paint droplets all over the metal surface. The advantages are firstly, that the metal will get an even coat, as the paint will spread out evenly all over the metal surface, secondly the paint will be attracted into inaccessible corners, and thirdly, there will be less wasted by falling on the ground or hitting surrounding walls. If there is less wasted this will save money.

Topic 7 Radioactivity

Atomic Structure
Test Yourself
1 **a)** Protons, neutrons, electrons; **b)** Protons, neutrons.
2 There are 8 protons in the nucleus (and 8 orbital electrons when the atom is neutral).
3 nitrogen-14 and nitrogen-16.

Stretch Yourself
1 $^{235}_{92}\text{U} \rightarrow {}^{231}_{90}\text{Th} + {}^{4}_{2}\text{He}$
2 $^{16}_{7}\text{N} \rightarrow {}^{16}_{8}\text{O} + {}^{0}_{-1}\text{e}$

Radioactive Decay

Test Yourself
1 a) Alpha particle; b) Alpha particles and beta particles.
2 a) ½; b) 1/16

Stretch Yourself
1 A helium atom.
2 a) $16000 \div 2 = 8000$ Bq; b) $16000 \div 2^3 = 2000$ Bq;
c) $16000 \div 2^{10} = 16$; d) 16000 Bq $\times \frac{1}{2} \times \frac{1}{2} \times \frac{1}{2} \times \frac{1}{2} \times \frac{1}{2} = 500$
Bq = 5 half-lives = $5 \times 3 = 15$h

Living with Radioactivity

Test Yourself
1 50% (or half)
2 From kill, damage, damage DNA, turn cancerous.
3 Her risk of getting cancer is increased, but we can't tell whether she will get cancer.

Stretch Yourself
1 Cornwall and Cumbria or other red areas on the map.

Uses of Radioactive Materials

Test Yourself
1 a) e.g. smoke detector; b) e.g. medical tracer; c) e.g. cancer treatment.
2 Each of the beams is a low dose so it doesn't kill the cells, but when combined in the tumour the dose is high enough to kill the cells.

Stretch Yourself
1 Alpha is stopped by the smoke, beta and gamma are not.
2 There is insufficient radioactive carbon left in the egg, so it could be much older – you can't tell.

Nuclear Fission and Fusion

Test Yourself
1 Fission is splitting large nuclei into two roughly equal parts, fusion is joining two small nuclei.
2 The plutonium-239 nucleus absorbs a neutron splits into two parts and a few extra neutrons. These neutrons are absorbed by more plutonium-239 nuclei which split – and so on.
3 To absorb neutrons and stop them causing more nuclei to fission. The rods are lowered or raised to change the number of neutrons absorbed and control the rate of reaction.
4 Stored under water in cooling tanks (high level waste).

Stretch Yourself
1 0.1 g $= 0.0001$ kg $\times (3 \times 10^8$ m/s$)^2 = 9 \times 10^{12}$ J
2 The strong force.

Practice Questions
1 D
2 $\frac{1}{16}$; 3 B
4 a) one sixteenth; b) i) It would decay away before reaching the organ and being recorded by the gamma camera; ii) the patient would stay radioactive for days.
5 a) They fired alpha particles at thin gold foil in a vacuum and counted the flashes of light from the alpha particles striking the fluorescent screen.
b) Most alpha particles went straight through. Some were deflected. A very few came straight back from the foil. c) The atom was mostly empty space. Mass was concentrated in a small volume and had a positive charge.
6 a) Sodium-24 has one extra neutron; b) 15 hours; c) 20 counts per minute; d) i) No, because it would decay too fast and there would be no radioactivity left after a few weeks/because it is a gamma emitter and people who came close to it would be irradiated; ii) Yes, because it is a gamma emitter so the radiation from the pipe would reach the surface/ because it has a short enough half-life not to contaminate the site for a long time
7 You will be given an injection containing radioactive iodine-131. The iodine will be absorbed by your thyroid gland. The radioactive iodine nuclei will decay by emitting beta particles. Beta particles are ionising and they can kill living cells, so many of the beta particles will kill the tumour cells. In 8 days half of the iodine-131 will have gone and in 16 days there will be only a quarter left. After 80 days, which is 10 half-lives, there will be less than one thousandth left. To decide whether to have the treatment you must weigh up the benefits and the risks. The risk is that beta particles will hit healthy cells and kill or damage them. You can replace a few killed cells, but a damaged cell could mutate and cause a new cancer. This doesn't mean it will cause cancer, only that there is a risk. The benefit is that the beta particles will kill the tumour cells so that you are cured.

Topic 8 Light

Refraction, Dispersion and TIR

Test Yourself
1 It gets faster.
2 Violet
3 e.g. optical fibres used in endoscopes.
4 One to transmit light and one to transmit the image.

Stretch Yourself
1 $\sin 20° \div \sin 8° = 2.46$
2 $\sin c = 1 \div 1.5$; $c = 41$–$8°$

Lenses

Test Yourself
1 Lens A
2 Fatter, it will have a shorter focal length.

Stretch Yourself
1 $1 \div 4 = 0.25$ m or 25 cm
2 6 cm $\div 2.4$ cm $= 2.5$

Seeing Images

Test Yourself
1 Ciliary muscles.
2 Converging/convex.
3 Diverging
4 Virtual, upright, magnified, behind the object.

Stretch Yourself
1 It would make it more curved.
2 Upright, virtual, smaller than the object.

Telescopes and Astronomy

Test Yourself
1 Eyepiece = A; objective = B.
2 5 cm – will collect more light.
3 A large concave mirror.
4 No

Stretch Yourself
1 1.5
2 The one with angle 1".

Practice Questions
1 D
2 $\sin r = \sin 40° \div 1.5 = 25°$
3 a) 3×10^8 m/s $\div 1.33 = 2.3 \times 10^8$ m/s; b) $\sin^{-1}[1 \div 1.33] = 49°$
4 Focal length of the lens; height of the object; distance of the object from the lens.
5 5 mm $\times 4 = 20$ mm
6 a) Away from bulb; b) Projector
7 Lens focuses colours at different points, difficult to make large lenses, heavy, uneven glass, but mirrors can be supported at the back.
8 a) Suspensory ligaments; b) The eye lens gets fatter; c) i) On the retina ii) In front of the retina; d) A diverging lens.
9 a) $0.5 \times 1" = 0.5"$; b) 1 parsec is distance to star with parallax angle of 1", a star with angle 0.5" is twice as far away, so 2 parsecs.
10. The eye and a digital camera both have a small hole for the light rays to pass through, called the pupil in the eye and the aperture in the camera. The iris changes the size of the pupil to let more or less light in, whereas the camera has an iris diaphragm to change the aperture. The light rays are focussed by a converging lens in both cases and form a small, real, upside down image on the light sensitive surface, which is the retina in the eye, and the CCDs (charge coupled devices) in the digital camera. These produce electric signals which travel along the optic nerve to be stored by the brain in the eye, and along wires to be stored on the memory card in the camera. To change the distance at which the eye focuses the ciliary muscles change the shape of the lens, but in the camera the lens is moved to change the distance between the lens and the screen.

Topic 9 Further Physics

Electromagnetic Effects
Test Yourself
1 Clockwise
2 The voltage would be reversed.
3 It increases it.

Stretch Yourself
1 920×12 V $\div 230$ V $= 48$
2 230 V $\times 96 \div 920 = 24$ V

Kinetic Theory
Test Yourself
1 They gain energy and break the bonds holding them together as a solid, so ice melts and becomes liquid water. They continue to gain energy and their average KE increases as temperature rises.
2 4000 N $\div 2$ m^2 $= 2000$ Pa
3 $100 + 273 = 373$ K
4 $195 - 273 = -78$°C

Stretch Yourself
1 They are making more collisions with the surface of the container.
2 Particles of gas B have twice the average KE of particles of gas A.

The Gas Laws
Test Yourself
1 Pressure halves.
2 Volume is three times.
3 Temperature in K is halved.
4 They stop moving, have no kinetic energy.

Stretch Yourself
1 $T_1 = 30 + 273 = 303$ K; $T_2 = 21 + 273 = 294$ K
$P_2 = \dfrac{6 \times 10^5 \text{ Pa} \times 294 \text{ K}}{303 \text{ K}} = 5.8 \times 10^5$ Pa
2 $T_2 = \dfrac{1.2 \times 10^5 \text{ Pa} \times 0.8 \text{ m}^3 \times 290 \text{ K}}{1 \times 10^5 \text{ Pa} \times 0.5 \text{ m}^3} = 557$ K

Medical Physics
Test Yourself
1 Sound with frequency greater than 20 kHz.
2 d = ½ × 1500m/s × 0.08s = 60 m
3 a) Proton, positron; b) positron, electron

Stretch Yourself
1 Time = 5×20 μs; ½ time = 50 μs; d = 1500 m/s $\times 50$ μs $= 0.075$ m
= 7.5 cm (or 4 divisions = 6 cm so 5 = 6 ÷ 4 × 5 = 7.5 cm)
2 $_1^1\text{H} \rightarrow\ _0^1\text{n} +\ _{-1}^0\text{e}$

Practice Questions
1 B E D C A
2 a) It will increase in one direction, reduce to zero, increase in the other direction and reduce to zero again; b) Electromagnetic induction; c) The voltage will change more quickly and it will be larger.
3 a) Because the voltage is not changing and there must be motion to induce a voltage; b) The voltage and current is changing which produces a magnetic field in the core, and the output coil. The output coil is in a changing magnetic field so a voltage is induced; c) $N = 200 \times 12$ V $\div 3$ V $= 800$ turns; d) Step-up has more turns in secondary coil than the primary coil, step-down has fewer.
4 a) N to S; b) i) yes (down) ii) no iii) yes (up) iv) no; c) Anticlockwise; d) Change the magnetic fields, reverse the current; e) Increase the current, make the magnets stronger.
5 77 K
6 20°C = 20 + 273 = 293 K. 2 × 293 K = 586 K = 313°C
7 a) The speed of the particles increases so the collisions are more frequent and more force is exerted on each collision;
b) $P_2 = P_1 \times T_2 \div T_1 = 3.5 \times 10^5$ Pa $\times 323$ K $\div 293$ K $= 3.9 \times 10^5$ Pa;
c) Cylinder would absorb energy from the Sun and this would raise the temperature which would increase the pressure and could cause it to explode.
8 A pulse of ultrasound waves is sent from a probe through the tissues of the mother's body. At each boundary between different tissue types some of the pulse is reflected. The reflections travel back to the probe and are detected. The time for the pulse to travel to the boundary and back is given by: time = 2 × depth of boundary ÷ ultrasound wave speed. From the times that all the reflected pulses take to arrive, a computer can build up a 3d image of the fetus. The advantages of ultrasound scans are that ultrasound can detect boundaries between different soft tissues, which X-rays can't, and that ultrasound is not ionising radiation, so it is not harmful to cells, unlike X-rays which are ionising and can damage DNA, leading to a risk of cancer.

Answers

Periodic Table

Key

| relative atomic mass |
| **atomic symbol** |
| name |
| atomic (proton) number |

| | | 1 | | | | | | | | | | | H | hydrogen | 1 |

1	2											3	4	5	6	7	0
																	4 **He** helium 2
7 **Li** lithium 3	9 **Be** beryllium 4											11 **B** boron 5	12 **C** carbon 6	14 **N** nitrogen 7	16 **O** oxygen 8	19 **F** fluorine 9	20 **Ne** neon 10
23 **Na** sodium 11	24 **Mg** magnesium 12											27 **Al** aluminium 13	28 **Si** silicon 14	31 **P** phosphorus 15	32 **S** sulfur 16	35.5 **Cl** chlorine 17	40 **Ar** argon 18
39 **K** potassium 19	40 **Ca** calcium 20	45 **Sc** scandium 21	48 **Ti** titanium 22	51 **V** vanadium 23	52 **Cr** chromium 24	55 **Mn** manganese 25	56 **Fe** iron 26	59 **Co** cobalt 27	59 **Ni** nickel 28	63.5 **Cu** copper 29	65 **Zn** zinc 30	70 **Ga** gallium 31	73 **Ge** germanium 32	75 **As** arsenic 33	79 **Se** selenium 34	80 **Br** bromine 35	84 **Kr** krypton 36
85 **Rb** rubidium 37	88 **Sr** strontium 38	89 **Y** yttrium 39	91 **Zr** zirconium 40	93 **Nb** niobium 41	96 **Mo** molybdenum 42	[98] **Tc** technetium 43	101 **Ru** ruthenium 44	103 **Rh** rhodium 45	106 **Pd** palladium 46	108 **Ag** silver 47	112 **Cd** cadmium 48	115 **In** indium 49	119 **Sn** tin 50	122 **Sb** antimony 51	128 **Te** tellurium 52	127 **I** iodine 53	131 **Xe** xenon 54
133 **Cs** caesium 55	137 **Ba** barium 56	139 **La*** lanthanum 57	178 **Hf** hafnium 72	181 **Ta** tantalum 73	184 **W** tungsten 74	186 **Re** rhenium 75	190 **Os** osmium 76	192 **Ir** iridium 77	195 **Pt** platinum 78	197 **Au** gold 79	201 **Hg** mercury 80	204 **Tl** thallium 81	207 **Pb** lead 82	209 **Bi** bismuth 83	[209] **Po** polonium 84	[210] **At** astatine 85	[222] **Rn** radon 86
[223] **Fr** francium 87	[226] **Ra** radium 88	[227] **Ac*** actinium 89	[261] **Rf** rutherfordium 104	[262] **Db** dubnium 105	[266] **Sg** seaborgium 106	[264] **Bh** bohrium 107	[277] **Hs** hassium 108	[268] **Mt** meitnerium 109	[271] **Ds** darmstadtium 110	[272] **Rg** roentgenium 111							

Elements with atomic numbers 112–116 have been reported but not fully authenticated

*The lanthanoids (atomic numbers 58–71) and the actinoids (atomic numbers 90–103) have been omitted.

The relative atomic masses of copper and chlorine have not been rounded to the nearest whole number.

Index

Index